BEGINNER'S GUIDE TO
AMATEUR ASTRONOMY

NORTH AMERICA NEBULA. Photo by Tony Hallas and Daphne Mount.

BEGINNER'S GUIDE TO
AMATEUR ASTRONOMY

An owner's manual for the night sky

David J. Eicher

KALMBACH BOOKS

Books by David J. Eicher

The Universe from Your Backyard; a guide to deep-sky objects from ASTRONOMY magazine
(Cambridge University Press and AstroMedia, New York, 1988)

Deep Sky Observing with Small Telescopes; a guide and reference
(Editor and coauthor; Enslow Publishers, Hillside, New Jersey, 1989)

Beyond the Solar System; 100 best deep-sky objects for amateur astronomers
(AstroMedia, Waukesha, Wisconsin, 1992)

Stars and Galaxies; ASTRONOMY's guide to observing the cosmos
(Editor and coauthor; AstroMedia, Waukesha, Wisconsin, 1992)

Galaxies and the Universe; an observing guide from Deep Sky *magazine*
(Editor; Kalmbach Books, Waukesha, Wisconsin, 1992)

The New Cosmos; the astronomy of our Galaxy and beyond
(Coauthor; Kalmbach Books, Waukesha, Wisconsin, 1992)

*Beginner's Guide to Amateur Astronomy; an owner's manual for the night sky from
the publishers of ASTRONOMY magazine* (Kalmbach Books, Waukesha, Wisconsin, 1993)

FOR MY WIFE LYNDA,
whose patience while I work at the keyboard is astronomical.

Cover design by Larry Luser
Cover photograph by Graham Sinagola

Printed in Hong Kong.

Eicher, David John, 1961-
    Beginner's guide to amateur astronomy : an owner's manual for the
night sky from the publishers of ASTRONOMY magazine / by David J.
Eicher.
       p.   cm. -- (Astronomy library ; no. 7)
    Includes bibliographical references and index.
    ISBN 0-913135-18-6
    1. Astronomy--Amateurs' manuals. 2. Astronomy--Observers'
manuals. I. Title. II. Series.
QB63.E36 1993
520--dc20   92-41663

Contents

Foreword

Every clear night as sunlight fades, the stars appear overhead and beam down their soft light. The Moon wheels across the sky each night, showing a cycle of phases and a pockmarked reminder of the violence of the early solar system. Shooting stars — meteors — flash overhead, leaving a faint trail of light that delicately vanishes. The luminescent arch of the Milky Way, our Galaxy viewed from within, towers overhead with its subtle beauty. Distant galaxies hide within the empty space above us, nonetheless brought into view by backyard telescopes.

The sky is rich in variety and gently shows us our place in the universe. Yet in this age of television, VCRs, camcorders, and computers, only a few people realize that the universe surrounds them or that Earth is a tiny, almost insignificant part of it.

Beginner's Guide to Amateur Astronomy is a call to anyone interested in learning about the universe. It introduces readers to the hobby of astronomy and shows how easily it can become an enjoyable part of their lives. Many think backyard astronomy is terribly technical. But typical of the book's low-key approach, the introductory chapter recommends that beginning skywatchers spend at least several months scanning the sky — and learning the constellations — armed only with a simple star chart and a pair of binoculars. Most beginning observers are surprised to learn that a pair of 7x50 binoculars shows dozens of interesting sky objects. They also provide a wide field of view ideal for learning your way around the sky.

Should you want a more permanent record of your observations later on, the chapter on imaging the cosmos describes how you can record various sky objects on film. Astrophotography is a relatively straightforward thing, at least in its simplest form. Load your camera with fast film, mount it atop a sturdy tripod, and open the shutter for 30 seconds to a minute. This will give you photos of some of the brighter constellations. Or expose for 15 to 30 minutes for a beautiful photo of "star trails," perhaps capturing a meteor as well. Closeup photos of the Moon, planets, and deep-sky objects can also be made with more elaborate techniques.

Observing with your own telescope is one of the great joys of amateur astronomy. Even a 2.4-inch telescope at low power, for example, will show you the rings of Saturn, belts on Jupiter, craters on the Moon, and the soft glow of a spiral galaxy. You'll see an amazing variety of stars, single, double, and multiple, many of which display colors. Observe on a dark, moonless night and you'll be able to see thousands of the stellar denizens of our Galaxy. Galactic deep-sky objects include star clusters — groups of stars held together by gravity — and fuzzy nebulae, regions of softly glowing gas and dust.

Whatever you view in the night sky, *Beginner's Guide to Amateur Astronomy* constitutes your owner's manual to the universe. Its author, Dave Eicher, has written a sweeping overview of the many magnificent aspects of astronomy available to amateurs, and it is a book you will keep on your shelf long after you are an old hand armed with a large telescope.

Welcome to amateur astronomy!

Robert Burnham
Editor, ASTRONOMY
Waukesha, Wisconsin
September 1993

Preface

Astronomy has come a long way over the last 400 years. In the seventeenth century the practice of astronomy was closely linked with astrology, and neither used anything like scientific thinking. The universe consisted of a "sphere" of stars and planets centered on Earth and no one had even imagined a universe that would include galaxies and clusters of galaxies stretching over billions of light-years. In 1609 the Italian scientist Galileo Galilei made a great step forward by training a telescope toward astronomical objects and discovering, among other things, the phases of Venus and Mercury, the moons of Jupiter, and the starlike nature of the Milky Way.

Such discoveries suggested a different structure for the heavens than the authorities had previously envisioned, and the Roman Catholic church promptly branded the great thinker a heretic. Indeed Giordano Bruno had been burned at the stake in Rome in 1600 for promoting the hypothesis that Earth orbits the Sun. The church did not murder Galileo, but they made his life pretty miserable. In 1992 the Pope officially admitted that Galileo was correct in hypothesizing a Sun-centered solar system and suggested that perhaps the church shouldn't have harrassed him. The change of heart took a mere 380 years.

Public opinion has been unkind to scientific thought often since Galileo's time. In 1665 Isaac Newton's pioneering experiments on light and gravitation were thought by many to be tinkerings of a madman. (Twenty-two years later Newton proved in his *Principia* the validity of the Sun-centered solar system.) In 1705 the English astronomer Edmond Halley predicted the return of a periodic comet. When the comet returned on cue in 1758, the public's skepticism faded. In 1796 Frenchman Pierre Simon Laplace described the nebular hypothesis of how the solar system formed. Despite the critical climate at the time, the accepted notion of how the solar system formed is essentially the same today.

As technology blossomed over the nineteenth century and was put to the test this century, humankind's view of the universe has changed in a revolutionary way. In 1918 the American astronomer Harlow Shapley first suggested the scale of the Milky Way Galaxy. Five years later Shapley's friend Edwin Hubble studied variable stars in the "spiral nebula" in Andromeda and found that galaxies are huge, independent systems outside our Galaxy and lying at vast distances. The breakthrough discoveries that defined the scale of the universe had been made.

We now live in the most exciting time ever for astronomy. With space-based and much larger ground-based observatories, improved technologies, and observations spread over all wavelengths, astronomers are poised to answer many of the fundamental questions about the universe over the coming years. How old is the universe? Will the universe expand forever or fall back on itself? How large is the universe? How many planetary systems exist within our local galactic neighborhood? Does intelligent life inhabit other planetary systems? The dawn of the twenty-first century is a terrific time to be involved with astronomy — it's much safer than the seventeenth century. And you can participate in these exciting times with a pair of binoculars or small telescope from your own backyard.

My thanks to my wife Lynda, whose support throughout this project has been strong and vigorous. I also extend thanks to ASTRONOMY's Richard Talcott, who cheerfully agreed to read the book and offered suggestions.

David J. Eicher
Waukesha, Wisconsin
September 1993

THE LAGOON AND TRIFID NEBULAE. Photo by Tony Hallas and Daphne Mount.

1

Cosmic Dust: An Introduction to Backyard Astronomy

Astronomy is like no other science. It can be enjoyed by anyone, at any time, who has the good fortune to have a clear sky overhead. Unlike archaeology astronomy's great laboratory is readily available to everyone, professionals and amateurs alike. And contrary to popular notion, a pair of binoculars or a small telescope is quite capable of showing craters on the Moon, the moons of Jupiter, and many thousands of stars and galaxies. Hence, for a relatively small outlay of money and effort, it's possible to reach out beyond the fragile blue Earth and witness for yourself the universe that surrounds us.

Astronomy offers a journey of the imagination. We normally don't think of Earth being constantly bombarded by light from objects scattered all across the universe, but it is. All we need to do as astronomy enthusiasts is intercept some of that light to extend our vision from Earth's surface throughout our solar system, our Galaxy, and to galaxies and quasars beyond. Astronomy is not limited to the nighttime, either. Daytime astronomy, *a.k.a.* solar observing, affords an opportunity to view up close the star that powers

Earth and our solar system and makes life as we know it possible.

Before wading into this fascinating hobby, it's a good idea to review the basic structure of the universe. Our solar system is centered on the Sun, our local star, around which the planets orbit. The nine known planets in our solar system, in order of closeness to the Sun, are Mercury, Venus, Earth, Mars, Jupiter, Saturn, Uranus, Neptune, and Pluto. In addition to the planets, the solar system contains small rocky and icy debris that we see as comets, asteroids, and meteors. Distances in the solar system are measured using astronomical units (AUs). One AU is the distance between Earth and the Sun, approximately 93 million miles. Our star is but one in our Galaxy, the Milky Way, which contains several hundred billion stars. We exist in a spiral arm of the Galaxy about three quarters of the way out from the galactic center, in a relatively sleepy area, and slowly orbit the center. Our Galaxy's disk measures some 100,000 light-years across, a light-year being the distance light travels in one year (about 6 trillion miles). Yet even this gigantic scale is dwarfed by the immensity of the universe. Our Galaxy is but one of 100 billion or more that exist in groups and clusters spread across the expanding shell of the universe. Distant galaxies we can observe from Earth are 10 billion or more light-years away.

How to Get Started

Astronomy offers a broad spectrum of observational sights and activities. Millions of years ago hominids first gazed up in awe at the night sky. The twinkling points of light scattered across the heavens startled, confused, and fascinated ancient peoples like no other aspect of the natural world. Distant and mysterious, the stars seemed to symbolize something great and unknowable, attractive yet unreachable. Some of the stars seemed to twinkle; others moved relative to the fixed stars, as if directed by the gods. Time marched onward, and ages would pass before Earth's intelligent mammals would begin to understand the nature of the stars.

Yet in essence the stars, some of which we now know as planets, are sending us a message: that although we think of ourselves as somehow detached from them, they have played a significant role in making us what we

DUST AND GAS like that contained within the Orion Nebula long ago made possible the formation of stars, planets, and even life. Photo by Michael Stecker.

NIGHT FALLS over a large star party in southwestern Texas, and a flurry of activity on the ground is made visible by red flashlights carried by the observers. Photo by Tom Polakis.

are. Living beings are quite literally products of the stars — nothing more than cosmic dust organized in an incredibly complex way. Elements important to our chemistry were originally manufactured deep inside massive stars which eventually exploded as supernovae, seeding the Galaxy with the stuff of life.

Having been given such a fantastic entrance into the world and the surrounding universe, it's a sad testimonial that so few people are aware of their cosmic heritage. A great scientist once said that nearly 100 percent of humanity is born, lives, and dies on the little blue planet Earth without ever being aware of the immensity of the universe or their relationship to it.

Wouldn't it be great if a relaxing, enriching hobby allowed you to peer into the distant cosmos and better understand your place in it? In amateur astronomy you may discover that you enjoy casually scanning the skies with binoculars or laying back and watching meteors flash overhead. You may find that viewing the Moon and its thousands of rilles, craters, and valleys is spell-binding. You may attach yourself to the planets, closely observing the rings of Saturn, the ever-moving moons of Jupiter, or remote and challenging Pluto. You may become a deep-sky observer, finding fascination in hunting wispy nebulae and galaxies that lie millions of light-years away. You may be a work-shop astronomer who grinds mirrors and builds finely crafted telescopes. You may fall in love with astrophotography and collect your personal set of astro images. Or you may want to casually read about astronomy and what scientists are finding out about cosmology without ever venturing outside to observe.

Naked-Eye Astronomy

Whatever path you choose, astronomy is a large and active hobby that has room for every kind of enthusiast, demands little, and offers a lifetime of fascination. The best way to begin your enjoyment of the night sky is to get a simple star chart (the center spread map in ASTRONOMY Magazine is ideal), head outdoors on a clear, moonless night and learn a few constellations (which are arbitrary patterns of stars created by ancient cultures to, in many cases, honor mythical figures). This requires nothing more than some warm dress, a mug of coffee or tea, and a little patience. Your instrument will be your eyes alone, which under a country sky will reveal thousands of stars, some as faint as 6th magnitude. (The system of magnitudes provides astronomers with a measure of visual brightness, with larger numbers representing fainter objects. The planet Jupiter, for example, normally shines at around -2 magnitude, Mars at magnitude 1, and the brightest stars in the Little Dipper at magnitude 3. The faintest stars visible with the naked eye are about 6th magnitude, the faintest stars visible in most backyard telescopes about 15th magnitude.) Pick out some commanding star groups to focus on. In the summer, Sagittarius, Scorpius, and Cygnus are favorites. In autumn check out the Great Square of Pegasus. During winter you can view Orion and Taurus. As spring thaws the landscape, constellations like Ursa Major, Draco, and Virgo will greet stargazers in the evening sky.

SATURN'S RINGS capture the imagination of everyone who views them through a telescope. NASA photo.

As you learn the bright constellations, you can branch out into identifying some fainter ones. These include Lacerta (summer), Pisces (autumn), Eridanus (winter), and Cancer (spring). As your knowledge of the sky grows, your ability to find unusual objects — stars, star clusters, and galaxies — will allow you to show friends distant inhabitants of the universe should you one day get a telescope.

Don't neglect the planets, either, which move in the sky relative to the stars as they orbit the Sun. Brilliant Venus — the "morning star" or "evening star" — is unmistakable, though it never strays terrifically far from the Sun. Fainter, ruddier Mercury is visible to the naked eye as well, and like Venus remains close to the Sun in our sky. Bright planets that traverse the path of the

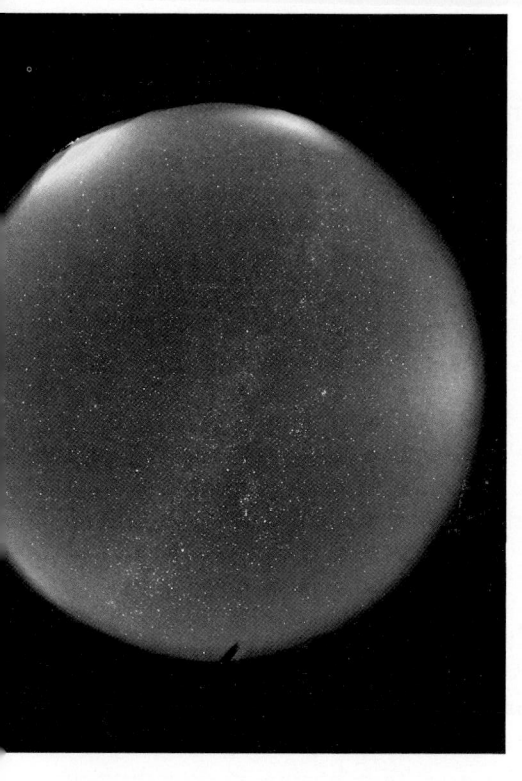

A SNAPSHOT OF THE SKY captures the winter constellations Orion, Taurus, and Canis Major, and includes the faint band of light called the Milky Way. Photo by Michael Terenzoni.

THE CRATER COPERNICUS is one of the thousands of features on the Moon's surface visible in small telescopes. Photo by Gerard Therin.

ecliptic — the plane of our solar system as we see it on the sky — include Mars, Jupiter, and Saturn. Their motion is slow enough that you can effortlessly follow their changes against the background stars over days, weeks, and months. Uranus is barely visible to the unaided eye under a dark sky, so picking it out among the stars is an invigorating naked-eye challenge. The two most distant planets, Neptune and Pluto, are beyond the reach of human eyes alone. To enter the magnified world of the cosmos, you'll need a good pair of binoculars. A standard, off-the-shelf set of 7x50s will do nicely.

Astronomy with Binoculars

By providing a jump in magnification and the added light-gathering power of a lens, binoculars bring you much closer to a wide array of astronomical objects. One of the most magical eras of my astronomical observing is that period of nearly a year when I roamed the sky armed with only binoculars. Without the experience and the telescope that came later, I had little idea of what various objects might look like. Yet nearly every clear night I carried outside a simple star chart and my pair of binoculars and scanned the stellar veil of the Milky Way, the barren fields of the spring and autumn skies, and the frosty, bright constellations of winter. Every star and every fuzzy object brought new fascination and a new mystery to explore.

A Backyard Observatory

At some point, you're going to want to make the jump to a telescope. This marvelously simple instrument consists of mirrors and/or lenses set in a tube. Three basic types of telescopes exist. The simplest is a Newtonian reflector, a telescope that employs a primary mirror to focus an image with the help of an eyepiece. Simple reflectors offer exceptional value by providing the biggest possible mirror — and therefore the most light gathering power — per dollar. A good 6-inch reflector will show all of the planets nicely and approximately 10,000 deep-sky objects — star clusters, nebulae, and galaxies. As with any telescope, the larger the mirror, the better. The limiting factors are expense and portability. At some point a telescope stops being a device that you can easily carry around.

The second major telescope type is a refractor, a lens-based instrument based on the original design used by Galileo in 1610. Refractors provide razor-sharp images and therefore are particularly good for observing planets and other small objects where seeing detail is critical. However, refractors are relatively expensive inch for inch and large refractors are just that — physically very large.

Catadioptric telescopes constitute the third major type and consist of lens/mirror combinations. They are extremely portable because they employ

HALLEY'S COMET (lower right) sports a short tail during its last close pass near the Sun in March 1986. The patch of light near the center of the photo marks the direction of the center of our Galaxy. Photo by Gordon Garradd.

THE PELICAN NEBULA in Cygnus is one of thousands of faint nebulae visible in backyard telescopes. Photo by Robert W. Provin and Brad D. Wallis.

optics in a folded light path so their tubes can be kept short. They are moderately expensive, but offer superb qualities for backyard observers.

Exploring the Solar System

Okay, you've got a telescope. Now what are you going to look at? The obvious first choice is to scope out the solar system, our bright and easily observed group of celestial neighbors. The brightest of them all is of course the Sun, the star that powers our bustling activities on Earth. Solar observing means setting up your telescope during the day when it's easy to see how to operate the telescope; therefore, it provides a good training ground for operating one. However, it is potentially dangerous. Never observe the Sun without an attached solar filter on the front (aperture) end of your telescope. Such filters remove powerful radiation from the focused sunlight that could otherwise blind you in a split second. With a proper solar observing setup, however, you can observe sunspots, flares, and other phenomena at your leisure.

The most popular object for beginning astronomers is the Moon. Speckled with mare, a smooth dark area, and pocked by craters, the lunar surface offers hours of exploration with even the smallest telescopes. The range of detail is so great and the Moon so different in appearance during its cycle of phases that you may find yourself hooked on lunar observing for a lifetime.

At least once in a while, however, you'll want to explore the planets. The "big three" from a telescopic perspective are Mars, Jupiter, and Saturn. Particularly when it is relatively close to Earth, Mars' orange-colored disk shows a surprising amount of albedo features — that is, dark regions visible against the planet's brighter surface. Jupiter's mighty bands, belts, and Great Red Spot show easily in small scopes, as does its ever-present Galilean moons, which comprise a miniature solar system revolving about the mighty planet. Saturn is the greatest spectacle of the solar system. I'll never forget the night in 1975 when I first glimpsed Saturn's rings through a small reflecting telescope. The planet hovered like a magic illusion, and I was simply awestruck that the sight was so easily visible from a friend's backyard.

MIGHTY JUPITER, largest of the planets, displays a complex system of belts and zones in backyard telescopes, as well as the mysterious storm known as the Great Red Spot. NASA photo.

THE STARS OF CASSIOPEIA trail behind an aged tree in this dramatic "nightscape." Such astrophotos are simple to make and require only a camera, tripod, and film. Photo by Dick Suiter.

Other planets are visible, too. Venus and Mercury show a cycle of phases similar to the Moon's as they orbit the Sun. Uranus and Neptune are challenging, distant worlds, visible in telescopes as nonstellar disks with blue-green hues. Pluto, the last planet of the solar system, is so small and so remote it requires an 8-inch telescope to be seen and even then it appears only as an extremely faint "star."

When observing planets, you'll notice that compared with many sky objects they are very small. Astronomers measure the sizes of objects in the sky using degrees, arcminutes, and arcseconds. One degree (abbreviation: °) contains sixty arcminutes (abbreviation: '), and one arcminute contains sixty arcseconds (abbreviation: "). The Full Moon, for example, is about one-half degree, or 30', across. Many galaxies and nebulae span 10' or so. Planets, however, are much smaller: the disk of Mars is typically less than 10" in diameter.

To the Depths of the Universe

Sun, Moon, and planets do not a universe make, however. The sky is also filled with billions of stars belonging to our Galaxy, the Milky Way, and many peculiar inhabitants known as deep-sky objects. Many of these interesting telescopic objects are double, multiple, or variable stars — star systems consisting of more than one sun or systems that vary in light output for a handful of reasons. Other deep-sky objects are more exotic residents of the Milky Way. Star clusters are groups of suns born together from a common cloud of dust and gas. These groups of stars slowly revolve about the center of the Galaxy until they break apart and travel separate stellar paths. Nebulae are gas and/or dust clouds of several distinct types. Bright nebulae represent the extremes of stellar life, generally being gas clouds associated with star birth or death. Dark nebulae are scattered clouds of dust floating throughout the Galaxy whose presence is known only when they block light from objects beyond.

Numerically, however, another type of deep-sky object outdoes all others in the entire sky. Thus far everything we've surveyed has been a component of our solar system or our Galaxy. Yet the universe is composed of countless billions of galaxies, arranged in groups, clusters, and superclusters. Thousands of galaxies are visible with a backyard telescope, as are a small number of their highly energetic cousins, quasars.

In observing deep-sky objects, one technique is particularly useful. Averted vision, glancing to one side of the eyepiece field, lets you use the most sensitive part of your eyes — the rods. These receptors are best at detecting faint light, so the technique is especially good at letting you see the most detail in faint objects.

Astronomical Photography

You're hardly limited to observing the universe. Many amateurs collect their own images of the sights they have seen in the solar system, Galaxy, and beyond. Astronomical photography is easier than it sounds, and is becoming easier every day as improved cameras and films hit the market. The simplest way to start is with camera and tripod, taking photos of star trails, meteors, and nightscapes. This entails simply setting up the camera, loading it with fast film, composing a shot that includes some horizon landscape detail as well as sky, and opening the shutter for several minutes. The results will surprise you.

Of course more sophisticated astrophotography is possible as well.

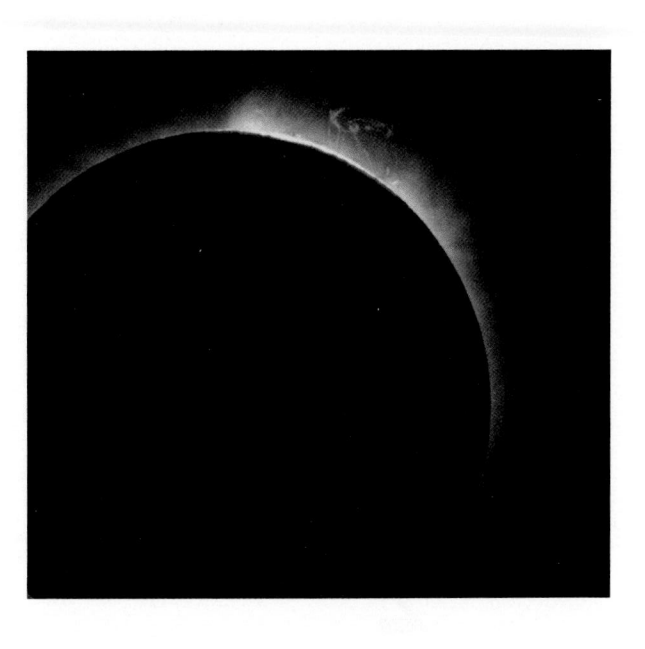

A MASSIVE PROMINENCE leaps from the Sun's limb during the solar eclipse in July 1991. The size of the prominence dwarfs the diameter of Earth. Photo by Rajiv Gupta.

Guiding the camera so it tracks the motion of the stars across the sky produces stars-as-dots rather than stars-as-streaks results. It's possible to take guided portraits of constellations, of deep-sky objects, and of the Moon, Sun, and planets. Ultrasophisticated electronics are also invading the world of astronomical imaging. CCD chips and computers are taking their places alongside cameras and darkrooms. High-end astroimaging is possible, and we will look at the possibilities it offers.

A Walk Through the Universe

But before you explore all the options, enter the hobby of astronomy at a slow walk. If an astronomy club exists in your area, drop in for a meeting or two and get to know the flavor of the hobby. (Appendix 4 on page 154 lists each of the major astronomy clubs in the U.S. and Canada.) Subscribe to the leading magazine of the astronomical hobby, ASTRONOMY. If possible, attend a few star parties to glimpse distant worlds through the eyepiece of an amateur's telescope. Try different types of equipment and see what you like best. Make sure you observe from the darkest possible site, away from the light pollution of city and suburban skies. Astronomy is a hobby that will only expand as man's conciousness of the universe expands. Most importantly, approach your exploration of the universe with an open mind. In his 1981 work *Cosmic Discovery*, Harvard astrophysicist Martin Harwit predicted that astronomers have identified to date only 10 to 20 percent of the phenomena that exist in the universe. What an exciting, unpredictable future lies ahead!

2

From Naked Eye to Observatory: Equipment for Skygazing

One of the greatest aspects of astronomy is that it can be enjoyed with equipment we all have — eyes. A stroll under the stars on a dark night will demonstrate the ability of the eye. Its resolving power and dynamic range to show bright and faint objects is amazing. It can show you the brightest astronomical object, the Sun, on a clear afternoon. (You should never look at the Sun without a proper filter, however.) It can show you all the planets except Neptune and Pluto. It can show you meteors, comets, and aurorae. And it can show you thousands of stars down to about 6th magnitude. (The magnitude system, remember, is a measure of an object's brightness, assigning brighter objects smaller numbers.)

Though your eyes are your fundamental piece of equipment, they can

YOUR TELESCOPE is the key to unlocking the night sky. Telescopes range from department store toys to simple, effective reflectors to complex and expensive scientific instruments. Choosing the right one requires careful consideration. Photo courtesy Darla Gawelski/Kalmbach Publishing Co.

only show so much. To see faint astronomical objects, to see them up close, and to locate them, you'll want to use a pair of binoculars or a telescope and a good star atlas. You may also want to refer to the many superb books in the bibliography, which starts on page 162. Choosing the telescope or binoculars that is right for you takes patient consideration. The star atlas is somewhat more straightforward, however.

The first thing a star atlas should do is familiarize you with the constellations visible at the time you observe. To begin, check the Star Dome map in the center of each issue of ASTRONOMY. For a more detailed map that shows the positions of fainter stars and deep-sky objects, you may wish to use *The Observer's Sky Atlas* by Erich Karkoschka (Springer-Verlag, New York, 1990), *A Field Guide to the Stars and Planets* by Donald H. Menzel and Jay M. Pasachoff (Houghton-Mifflin, Boston, 1983), or the magnificently detailed *Sky Atlas 2000.0* by Wil Tirion (Cambridge University Press and Sky Publishing Corp., New York, 1981).

How to Choose a Pair of Binoculars

If you want to start looking at the sky with optics, the best way is with a good pair of binoculars. They'll let you explore the wonders of the sky with a wide field, scanning the summer Milky Way, viewing the bright star clusters of winter, and exploring the remote realm of the galaxies in spring and autumn. The experience you'll gain finding your way around the sky with just binoculars will be invaluable later with a telescope, and the time spent with binoculars will let you think about the attributes you'll look for in a telescope, a more major investment of both money and time.

Many beginning stargazers feel that binoculars can't possibly have the power to be useful in astronomy. Not true. As you'll see, magnification is not the most crucial aspect of your optical instrument. In fact, the low-magnification, wide fields provided by binoculars are a great help in seeing many objects. I recommend beginning with a pair of 7x50 binoculars — lenses that measure 50 millimeters across and magnify seven times — because they're big enough to show you stars, bright star clusters, and galaxies, but small enough to hold comfortably without a tripod.

A VARIETY OF TELESCOPE TYPES exist, each of which serves a different purpose. Don Sabers peers through a Newtonian reflector (right) while Marty Poissant checks the view through a 4-inch f/5 astrograph as Ronald E. Royer looks on. Photo courtesy Ronald E. Royer.

Moving to a Telescope

When you get the bug to buy a telescope, make sure you research the options first. No "best telescope" exists for all purposes, and the type of instrument you'll want to get will depend on your observational interests. Do you want to simply look at objects? Do you want to make long exposure, guided photographs of them? Are you predominantly interested in planets, double stars, and the Moon? Or do your tastes run toward star clusters, nebulae, and galaxies? Do you have a dark sky in your backyard or will you need to transport your scope to a good observing site? Such considerations weigh heavily toward the type of telescope you'll want.

Three basic classes of telescopes are commercially available, each sold by several manufacturers. Their advertisements can be seen regularly in the

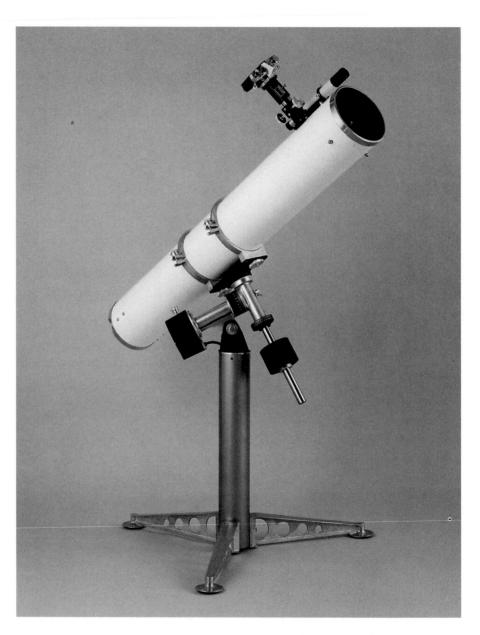

A NEWTONIAN REFLECTOR uses a primary mirror and a diagonal secondary mirror to reflect light into an eyepiece near the upper end of the tube. Photo courtesy Meade Instruments Corp.

pages of ASTRONOMY and other amateur astronomy magazines. The "classic" type of telescope is a refractor, with its iconic long, white tube and elegant tripod. This is the basic design used by Galileo in 1610. Refractors use lenses at the front (skyward) end of the telescope to focus light into an eyepiece that sits at the rear end of the tube. Achromatic (two-lens) designs have been available for years and are relatively good buys, but they produce minor aberrations — color distortions — in the images they deliver. Apochromatic (three-element lenses) refractors provide razor-sharp images of planets and pinpoint stars. They typically have long focal lengths, meaning they provide relatively small

THE ORIGINAL TELESCOPE DESIGN, a refractor uses a lens at one end of a long tube to bring light to a focus at the other end. Jim Phillips and Tom Dobbins inspect a newly finished 8-inch f/13 refractor. Photo courtesy Jim Phillips.

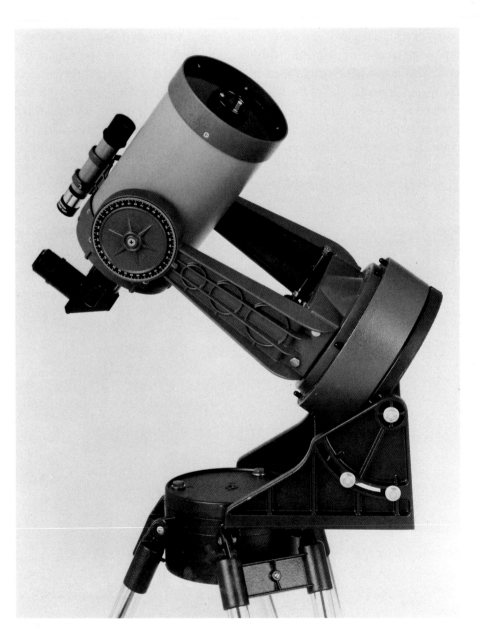

A SCHMIDT-CASSEGRAIN TELESCOPE is one of several types of catadioptric designs that utilize both lenses and mirrors to provide large aperture in a compact package. Photo courtesy Celestron International.

DESIGNED AND BUILT FROM SCRATCH, Jean Dragesco's 4-inch folded solar refractor is optimized for viewing and photographing the Sun at high resolution. Photo courtesy Jean Dragesco.

TRAILER-MOUNTED FOR PORTABILITY, Canadian amateur Jack Newton's massive 20-inch reflector is set up for astrophotography under dark skies. Photo courtesy Jack Newton.

fields of view. Thus they are extremely good instruments for observing the Moon and planets and small objects like double stars and planetary nebulae. Recently manufacturers have produced shorter focal length apochromats that provide wider fields of view. These telescopes are optically superb, but the quality comes with a high price tag.

21

EQUIPPED WITH THE LATEST in high-tech accessories, California astrophotographer John Gleason's rig is capable of capturing on film thousands of deep-sky objects. Photo courtesy John Gleason.

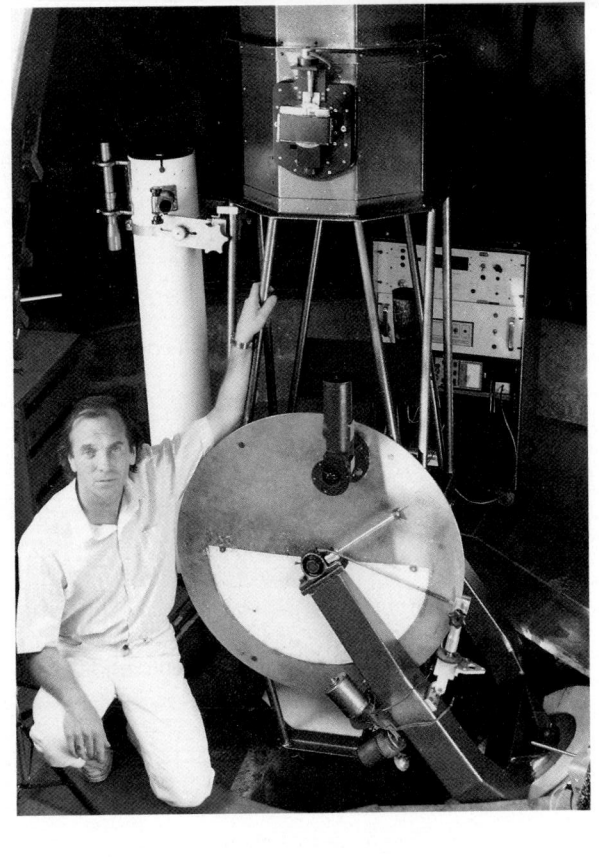

A SHEET METAL WONDER, English amateur Ron Arbour's telescope is finely machined and carries a full set of cameras, drive correctors, and electronic drive control accessories. Photo courtesy Ron Arbour.

Reflectors are the second major class of optical design. Modeled after the design used by Isaac Newton, these use a mirror in the rear of the telescope and a small secondary mirror suspended near the front of the tube. They bounce incoming starlight off the primary (rear) mirror, to the angled secondary, and into an eyepiece positioned on the side of the tube near the front end. These instruments typically possess shorter focal lengths than refractors and hence provide wider fields of view, making them very good telescopes for deep-sky observing. Their tubes are much shorter than those of refractors, making them more portable per inch of telescopic aperture.

The third major class of telescope is the catadioptric instrument, a combination of lenses and mirrors. The most popular type of catadioptric scope is the Schmidt-Cassegrain, which uses primary and secondary mirrors like a reflector but also employs a thin corrector plate at the front end. The design bounces the image back through the primary to an eyepiece at the telescope's rear. Because of the folded light path, Schmidt-Cassegrains pack relatively long focal lengths into a compact package, making them highly portable. A similar type of catadioptric is the Maksutov-Cassegrain telescope. All types of catadioptric telescopes are ideal for the Moon and planets and because of a vast array of ready-made accessories are particularly well suited for astrophotography.

The telescope is only part of the story. You need to get a selection of high-quality eyepieces if you expect your scope to perform well. A virtual

Telescope Types

ADVANTAGES	DISADVANTAGES
REFLECTORS	
Delivers color-free images Affordable in large sizes Wide field of view Eyepiece positioned at a comfortable spot	Secondary mirror obstructs part of light path Somewhat tricky to find objects Coma affects images at edge of field Long tubes somewhat unwieldy Requires collimation
REFRACTORS	
Least image defects of any design Typically superb optical quality Good lunar and planetary scopes Closed tube; no collimation needed	Relatively narrow fields of view Expensive Large apertures hard to use Eyepiece placed at awkward level
CATADIOPTRICS	
Ease of astrophotography Ease of portability Good light-gathering power Low-maintenance closed tube	Somewhat expensive Light path partly obstructed Slow focal ratios Narrow fields of view

A SIMPLE, HOMEBUILT PIPE MOUNT supports Australian amateur Peter F. Williams' reflector, with which he has made countless variable star brightness estimates. Photo courtesy Peter F. Williams.

explosion of high-quality eyepieces has surfaced over the past few years, and advertisements for them can be found in astronomy magazines. You'll also want to consider the simplicity of the telescope's mounting. If you absolutely do not want to do astrophotography, you can cut costs by getting a telescope on a so-called Dobsonian mount, named for John Dobson, who popularized them. A "Dob" is an altitude-azimuth mount equivalent to a battleship gun mount. It cannot track the stars in a single motion, and therefore you'll need to nudge the scope along to keep an object centered, but its simplicity and low cost will be assets if you are strictly a visual observer.

Before you choose a telescope for yourself, see if you can use a variety of telescopes at a local club star party. Read the annual telescope buyer's guide published in ASTRONOMY magazine. And talk to other amateurs to sound out their experiences. The most important thing to remember is that magnification — the classic advertising hype symbol — is not the important attribute of a telescope. For virtually all observing you'll be using powers of 40x to 150x. The important thing is the diameter of the lens or mirror, the telescope's aperture. You can see the Moon, planets, and hundreds of deep-sky objects with a 4-inch telescope. You can see thousands of deep-sky objects with an 8-inch telescope. As a general rule, you'll probably want to get the largest telescope you can afford — and lug around. Consider your observational interests and your interest in photography before you decide. It's a complex thing to determine, and a sufficient reason to spend a few months warming up with a good pair of binoculars.

3

Silver Halide and Electronic Chips: Imaging the Cosmos

Telescopes can be more than observational tools. They can also allow you to collect and store images of the sky objects you observe. Not only is it great fun to take photographs, collect electronic CCD frames, or simply sketch objects with a pencil and pad, but taken together, weeks, months, and years of collecting sky images leads to a personal inventory of your experiences as an amateur astronomer. It gives you plenty of good stuff to pull out of the closet on some cloudy night, a night when you can relive the fun of the past.

Keeping an Observational Journal

The easiest way to start your "imaging" of the sky is to keep a journal of what you observe. Your journal can be as simple as a pad of paper on which you write down what you've seen during each observing session, or as fancy as a buckram-bound book in which you note observations, make sketches, write in data for various objects you've seen, record comments on weather phenome-

THE SPLENDOR OF THE MILKY WAY is evident in this long exposure photograph of the center of the Galaxy. The bright star clouds in Sagittarius and Scorpius are visible to the naked eye. Photo by Ronald E. Royer.

THE PRECISE EXPOSURE enabled a skilled astrophotographer to capture the background stars during this 1979 lunar eclipse without overexposing the Moon itself. Photo by Jacques Guertin.

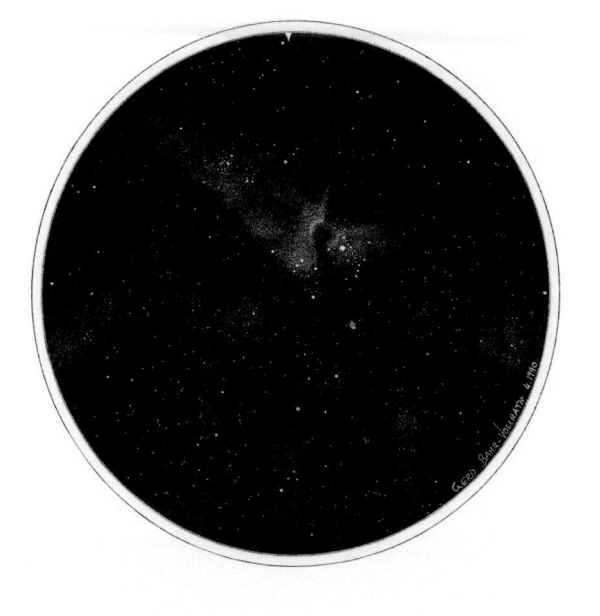

PENCIL AND ART PAPER created this magnificent sketch of the intricate Eta Carinae Nebula when coupled with the hands of a talented deep-sky observer and his 8-inch telescope. Sketch by Gerd Bahr-Vollrath.

na, and jot down anything that comes to mind about the universe at large.

Of all the ways to capture images of the cosmos, the easiest (and one of the most effective) is to sketch objects as your eyes see them. All this requires is a pad of paper and pencils. A No. 2 pencil works nicely, although an art pencil like an Eberhard Faber Ebony works best.

The technique is simple. First draw a circle that represents the telescope's field of view. Take the paper, pencil, and a dim red-filtered flashlight outside with your telescope. (The red flashlight will give you enough light to see what you're drawing but not so much that it ruins your eyes' adaptation to the darkness.) After you've set up and have an interesting object centered in the scope's field of view, simply look alternately into the eyepiece and down onto the paper. Carefully plot the brightest stars in the field on the drawing. Use geometric patterns and distances to gauge the placements of these stars in your circle on the paper. Now add the fainter stars, using the bright stars as guideposts to place them correctly. Finally, you can rub a small amount of graphite from the pencil onto the paper and smooth it carefully with the light touch of a moist finger to simulate the fuzzy light from a galaxy or nebula. If you're looking at a planet, draw its disk first and then shade in any subtle details you see. It takes practice to get good at this, but sketching is inexpensive and relatively easy. Its most powerful benefit is one unmatched by any other form of imaging: it lets you record sky objects exactly as your eyes see them.

Astronomical Photography with Camera and Tripod

Another easy way to record your favorite astronomical objects is to photograph them. Although some forms of astrophotography are tricky, getting started is a joy. The equipment necessary for taking your first astrophotos consists of a camera capable of making time exposures (one with a shutter that can stay open for long periods), a lens of 28mm to 50mm focal length (the standard lens has a 50mm focal length), a roll of fast (generally 400 to 1600 ISO) film, and a sturdy tripod.

The best way to start is to set up before dark, having loaded film in the camera and attached it to the tripod. Carry the whole assembly outdoors and

CAMERA AND TRIPOD ASTROPHOTOGRAPHY is the simplest type yet produces some of the prettiest results. An unguided exposure of 15 seconds with a wide-angle-lens captured Venus, Mars, Saturn, and Mercury over the Golden Gate Bridge in San Francisco. Photo by Jim Baumgardt.

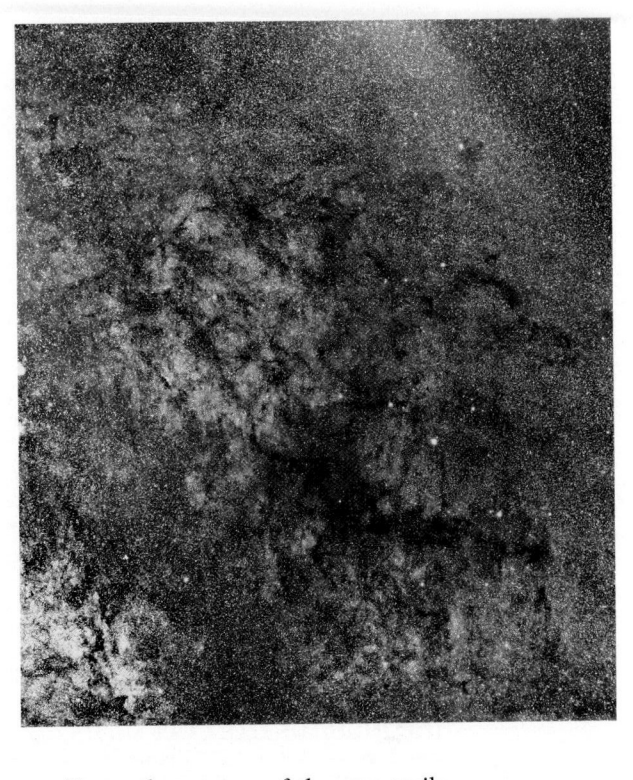

PIGGYBACK ASTROPHOTOS capture wide fields of sky by tracking a camera and lens piggybacked on a telescope used as a guiding platform. A 300mm lens and 240-minute exposure captured the dusty interstices between star clouds in the Milky Way. Photo by Brad D. Wallis and Robert W. Provin.

after nightfall aim it toward a favorite constellation — perhaps Orion in the winter or Sagittarius in the summer. Depending on your lens' focal length and where you are aiming in the sky, you can take photos of 15 to 30 seconds before Earth's rotation turns the stars into streaks. Take a number of these shots and you'll soon find yourself on the way to assembling a library of constellation photos. If you let the exposure go for longer — perhaps 10 to 30 minutes — you'll get pretty pictures of star trails. If you use color film, you'll see that many of the star trails show subtle blues, yellows, and oranges as well as stark white.

Camera-and-tripod photography is also great for capturing wide-field portraits of planetary conjunctions. You may find the Moon hanging in the sky near a bright planet. A 30-second exposure of the two during dusk will make a beautiful scene, especially if you compose the picture with an attractive foreground. The technique is also powerful for capturing bright aurorae.

Piggybacking Your Camera

If you have an equatorially mounted telescope with a clock drive to track the stars, you can go a step beyond camera-and-tripod photography and take "piggyback" astrophotos. With this technique you mount the camera and its lens atop your telescope and use the scope itself as a guiding platform to

THROUGH-THE-TELESCOPE SHOTS capture a great variety of closeups of solar system and deep-sky subjects. A diamond ring during the 1991 solar eclipse breaks through the clouds in this image. Photo by Jeff Schroeder.

move the camera in sync with the stars. The camera rides piggyback on the scope. No longer are you bound by short exposures; now you can take pictures of much fainter objects by keeping the shutter open for 15 to 30 minutes. If you properly align your telescope relative to the North Celestial Pole (see your camera's owner's manual), your piggybacked photos will contain stars that are small dots rather than being streaked into arcs. You may also use a special guiding eyepiece that contains an illuminated reticle — dimly lit crosshairs — so that you can watch a bright star in the telescope and make minute corrections in tracking during the exposure with a drive corrector.

Many subjects lend themselves especially well to piggyback astrophotography. You might use a wide-angle lens to capture vast panoramic views of the Milky Way's star clouds and nebulae. You might want to capture bright,

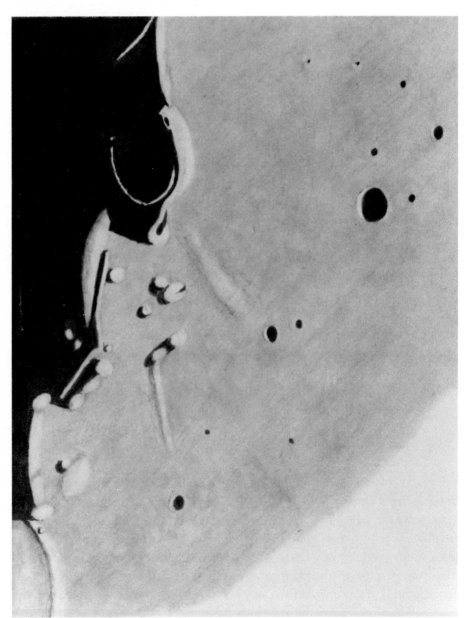

SUNRISE OVER THE CRATER MARIUS on the Moon's surface was captured by a skilled lunar observer equipped with sketch pad, pencil, and 6-inch refractor. Sketch by Jim Phillips.

large star clusters like the Pleiades and Hyades in Taurus using a 50mm lens, or capture long exposure shots of constellations that record faint stars and wisps of nebulosity. You may want to employ a telephoto lens of 150-300mm focal length to zero in on smaller deep-sky objects like individual galaxies and nebulae, capturing them in exposures of up to half an hour. The potential is unlimited, and experimentation with films, lenses, and exposure times is the best way to achieve consistently good results with your particular system.

Through-the-Telescope Photography

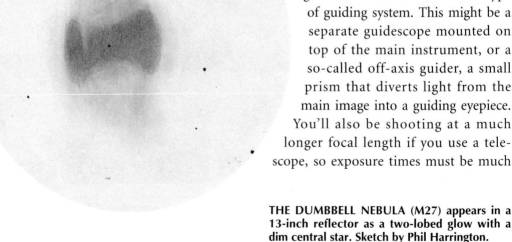

The trickiest type of astrophotography uses the telescope itself as a giant lens. Because the amount of sky seen in your telescope is so much smaller than with a camera lens, the result is a real closeup view of objects ranging from planets to galaxies. Sounds good? The tradeoff is that a multitude of variables such as focusing and guiding become more difficult when shooting through the scope. You now must attach the camera to the scope and focus it critically, guiding the exposure as it occurs, meaning you have to use a good drive corrector and some type of guiding system. This might be a separate guidescope mounted on top of the main instrument, or a so-called off-axis guider, a small prism that diverts light from the main image into a guiding eyepiece. You'll also be shooting at a much longer focal length if you use a telescope, so exposure times must be much

THE DUMBBELL NEBULA (M27) appears in a 13-inch reflector as a two-lobed glow with a dim central star. Sketch by Phil Harrington.

longer. Polar alignment is more critical at such long focal lengths. And finding objects is somewhat more difficult to begin with.

Simple astrophotography, particularly with camera and tripod, is the way to start. If after a few weeks or months of success you develop an appetite for taking pictures through your scope, begin by reading articles on the subject in ASTRONOMY. They'll provide the kind of detail you'll need to become a skilled astrophotographer behind the telescope.

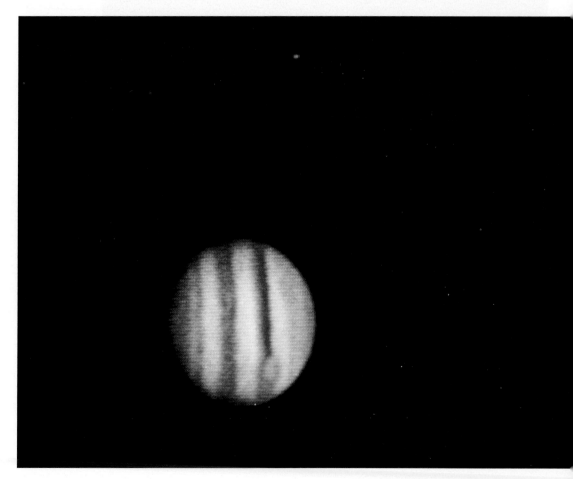

VIDEOGRAPHY with telescope and camcorder captured this image of Jupiter and its moon Ganymede in April 1990. Image by David Brewer.

THE MAGNIFICENT SPIRAL of galaxy M51 in Canes Venatici was recorded with an 11-inch telescope, hypersensitized Konica 1600 film, and a 70-minute guided exposure. Photo by Kim Zussman.

The Electronic Sky: Video Cameras, CCDs, and Computers

I can remember the "good old days," way back in the late 1980s, when the most complex and rewarding type of astroimaging was shooting photos through the telescope. Now a variety of electronic device makes it possible to capture beautiful images of planets, stars, and galaxies in digital form. Rather than taking the film down to the processor (or developing it in a tank in your bathroom), amateur astronomers often use their computers or television monitors as darkrooms. The kick here is that with a CCD camera you can take lots of images quickly — exposure times are often a fraction of those used with conventional films — and manipulate the image in the comfort of your study. You can also pick and choose between frames showing planets when the atmospheric seeing is momentarily rock steady, yielding razor-sharp images of Saturn, Mars, and Jupiter. You can also use image-processing software on your PC to maximize the brightness, contrast, color balance, filtration, or other aspect of your image, all instantly.

Video cameras are also useful tools for capturing astronomical images, although even with low-light-sensitive models the objects you shoot must be relatively bright.

Whether you start recording notes in a journal, take a few constellation photos, sketch what you see in the eyepiece, or get a CCD camera and open a digital darkroom, capturing the sights of the night sky is alluring. It offers great enjoyment for the future and the capability to share observations with your family or even a large group. Try a few experiments with astroimaging — you never know where they might lead you.

4

Dominant Spheres: The Sun and Moon

Two astronomical objects completely dominate the sky. One is so bright it blots out virtually everything else above the horizon. It is of course our own star, the Sun. The second object is Earth's Moon. Although it is much fainter than the Sun in terms of sheer brightness, the Moon — particularly at Full phase — is so bright that it seriously impairs seeing faint objects when it is up during the night. These two dominant spheres are the key objects in our sky; one because it is the central source of energy in the solar system and the other because it lies so close to Earth.

Our Own Star

Most beginning amateur astronomers think of astronomical observing as purely a nighttime activity. Of course most objects are visible only after dark, but it can be absorbing to observe the Sun itself. It's especially enthralling to observe the Sun if we think about what it means to us as earthlings.

Some 4.6 billion years ago the Sun condensed out of a cloud of inter-

THE SUN IN TOTAL ECLIPSE gives astronomers the opportunity to see the corona, the
Sun's tenuous atmosphere. Photo by Bill Sterne.

CRATERS, RILLES, VALLEYS, AND SEAS are visible on the Moon's surface when viewed with a small telescope. Photo by Rick Kurczewski.

stellar hydrogen gas and formed a sphere of high-density material. At a critical moment when enough mass was present the infant Sun blinked on, becoming a dazzling engine that fuses hydrogen atoms into helium. This process of nuclear fusion is the same in all stars, and it enables the Sun to produce the copious light and heat necessary for the sustenance of life on Earth.

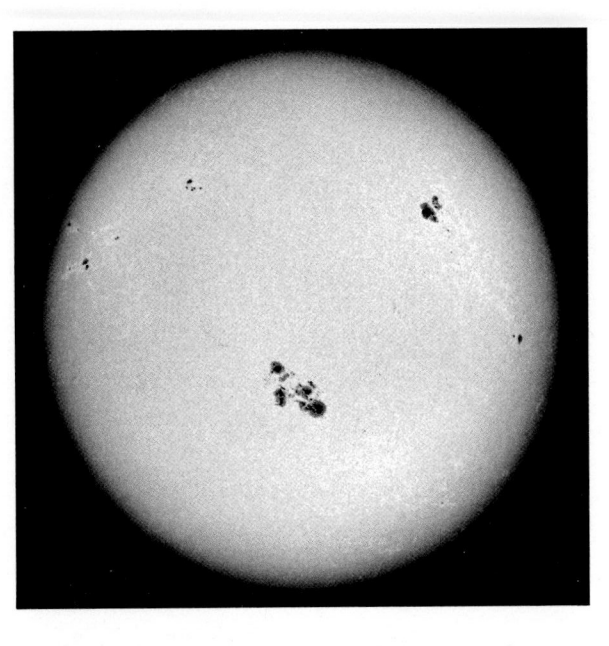

SUNSPOTS, magnetic storms on the Sun's surface, are routinely visible with a small telescope and Sun filter. Photo by Jean Dragesco.

Eons after the Sun ignited, uncountable chunks of debris from the formation of the solar system fell into orbits around it. Some of the inner material fell into the Sun and was vaporized. Much of the material some distance away from the Sun, primarily made of rock, gases, and ice, accreted into planets, asteroids, and comets. Chunks slammed into chunks and sent worlds careening off into orbits that eventually became stable. We are experiencing the net result of this event — minuscule as it is on a galactic scale — some 4.6 billion years after it began.

The Sun is average as stars go. Some stars are larger and hotter, others smaller and dimmer. However, because dim dwarfs outnumber all other stars, astronomers can say that the Sun is relatively impressive in terms of energy output. The Sun measures about 1,392,000 kilometers across at its equator — 108 Earths could be stacked side by side to fill up the distance. The Sun's mass is equivalent to 328,000 times that of the Earth-Moon system. The Sun is a G2 (yellow) star with an apparent magnitude of -26.8 and a surface temperature of 5800 K (Kelvin) — hot enough to melt steel into soup. Thankfully, it lies more than 93 million miles away, a distance astronomers call one astronomical unit (AU). Earth lies just in the right place for living beings: if the Sun were more distant or much closer, the temperatures and radiation would be either too cold and too weak or too warm and too strong for living things. Although our solar system may be a plain one set around a run-of-the-mill star, we're very fortunate to have it that way.

Daytime Astronomy: Observing the Sun

It's a joy to watch the behavior of the Sun on a warm spring or summer day, the sky a deep blue and the Sun's rays beaming down. A pair of binoculars or a telescope is capable of showing a variety of phenomena associated with the Sun. However, you must view it carefully and with the aid of a proper solar filter. Because of the intense radiation produced by the Sun, even a brief unfiltered look could cause irreversible damage to your eyes. So be careful and always use a solar filter.

There are several types of Sun filters. The best is an aluminized glass filter made to fit over the front end of the binoculars or telescope. These are fairly expensive; a more affordable alternative is an aluminized mylar-plastic filter that similarly fits over the front aperture of the binoculars or telescope. Several brands of these can be found in advertisements in astronomy magazines. Do not use solar filters that fit into telescope eyepieces. These are dangerous because they allow the intense radiation to enter the telescope full force and can crack or break suddenly, instantly blinding you. And that's not to mention the damage the intense heat can do to your scope's optics. Make sure the solar filter fits securely onto the telescope so no accident can happen, and be sure to cover unfiltered optics — like a finder telescope — that someone might mistakenly peek through. In addition to optical solar filters, several companies produce hydrogen-alpha solar viewers. These devices are not cheap but offer incredible views at the specific hydrogen-alpha wavelength of a multitude of solar activity.

Say you've set up your scope on a sunny day and want to do some Sun viewing. What are you going to look at? At first glance with the naked eye the Sun might seem a steady, unchanging disk of light. Actually, it is a dynamic, changing star with variable light output, sunspots, solar flares and prominences, and active regions called plages, faculae, and filaments. Moreover, on occasion the Moon passes in front of the Sun's disk and causes a solar eclipse, one of the most breathtaking sights in nature.

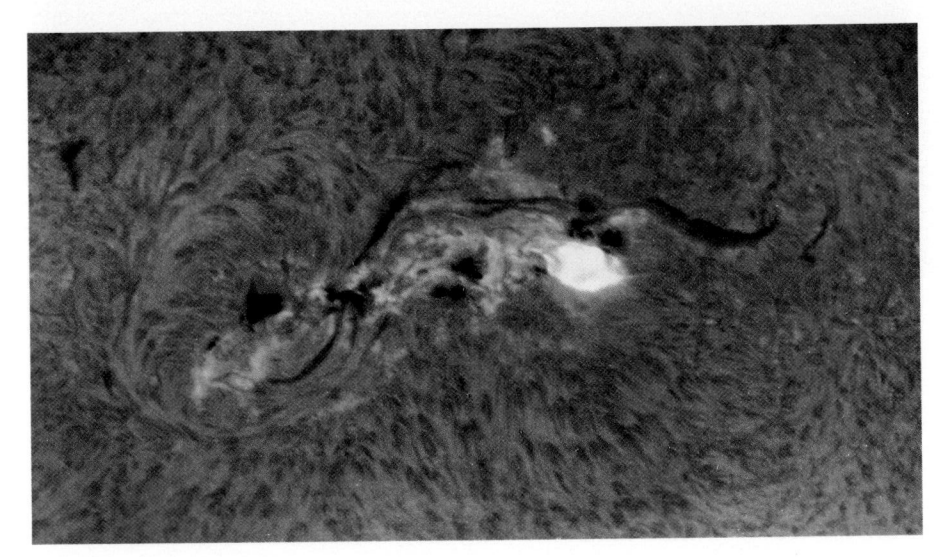

A LOOP PROMINENCE TOWERS above the Sun's limb. Such prominences are visible using a telescope specially fitted with a hydrogen-alpha filter. Photo by Jean Dragesco.

A SOLAR FLARE ERUPTS from a sunspot in this image made on September 4, 1989, in hydrogen-alpha light. Such flares send powerful bursts of energy outward, triggering aurorae on Earth. Photo by Jean Dragesco.

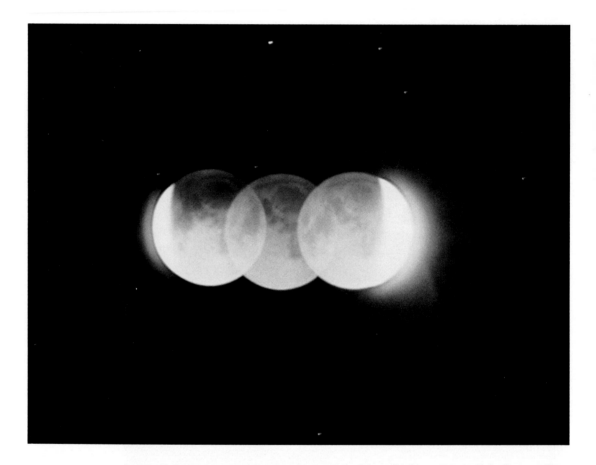

EARTH'S SHADOW CROSSES THE MOON in this multiple exposure of the lunar eclipse of May 6, 1982. Photo by Ronald E. Royer.

Sunspots are the easiest type of solar activity to view. The Sun undergoes an 11-year cycle of sunspots, so at times of great activity its surface is far more spotted than other times. But a filtered telescope aimed at the Sun more often than not shows a few sunspots, and occasionally they are large enough to be visible in binoculars or even with the unaided eye. Sunspots are magnetic storms on the Sun's surface, regions of cool temperature where a magnetic field breaks through the surface — the photosphere — and towers above, streaming radiation into space. Sunspots usually occur in groups, and are often preceded by faculae, whitish regions of superheated gas. An intensely active solar storm can produce a prominence, a loop or burst of gas that rises over the Sun's surface. These are best visible when they occur on the Sun's limb and with the use of a hydrogen-alpha viewer. Solar flares are energetic blasts of gas that may last for hours or only a few seconds. Although some are visible in white light, these too are best viewed at the hydrogen-alpha wavelength.

Check ASTRONOMY magazine for the coming eclipses of the Sun. You often have to travel some distance to see one, but it will stick with you for a lifetime. Three basic types of solar eclipses occur: total, partial, and annular. The differences simply reflect the orbital geometry of the particular time when an eclipse takes place. That eclipses occur as dramatically as they do is a freak accident in that the Moon's disk as we see it in the sky is almost exactly the same size as the Sun's. Hence, when the Moon passes in front of the Sun, it briefly covers it. A total eclipse is a "direct hit," when the entire Sun is blocked by the Moon. This enables astronomers to view the magnificently beautiful corona, the Sun's outer atmosphere, which is too faint normally to see. A partial eclipse happens when the Moon doesn't pass directly in front of the Sun. An annular eclipse occurs when the Moon is relatively distant in its orbit, covering the Sun but leaving a thin ring of bright sunlight around the solar circumference. Observe a solar eclipse or simply sunspots for several days and you might find yourself turning into a daytime astronomer.

A Battered, Barren World

An object of romance and wonder for ages, Earth's Moon is the dominant object of the nighttime sky (and is also visible frequently during the day). Because it lacks an atmosphere and active weather phenomena that resurface planets like Earth, the Moon is a stark contrast to the richness of our planet. The Moon's barren emptiness and heavily cratered terrain hark back to the early history of the solar system. Earth too has survived its heavy periods of bombardment from asteroids and meteorites (and Earth will be struck in the future, too!), but its resurfacing has hidden most of the scars of the past.

Collisions have played a major role in shaping the surfaces of planets, but in the case of Earth and the Moon, they played an even larger role. The

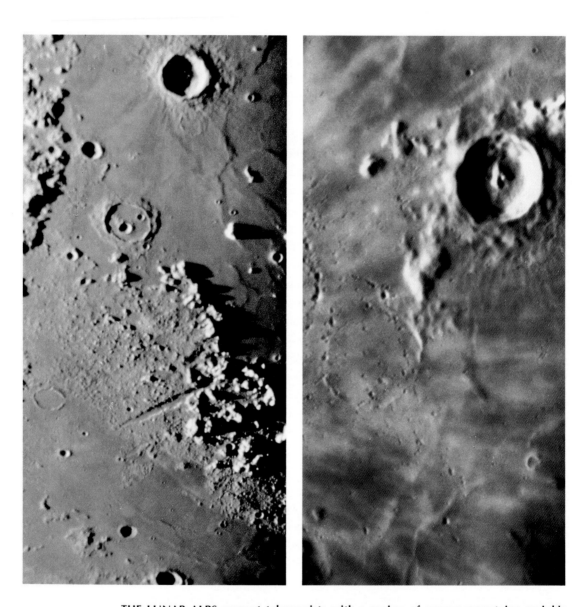

THE LUNAR ALPS present telescopists with a region of craggy mountains, wrinkle ridges, and small, deep craters. Photo by Jean Dragesco.

THE ANCIENT CRATER STADIUS, filled with lava, lies at the center of this high resolution image. Eratosthenes is the prominent crater at upper right; the chain of tiny craterlets at left was formed by the Copernicus impact. Photo by Gerard Therin.

leading hypothesis on the Moon's formation is that a huge collision between a Mars-sized body and Earth created the material that accreted into the Moon.

Most satellites in the solar system are relatively small compared with their parent planets. Yet the Moon is extremely large relative to Earth, making it a peculiar standout. The Moon consists of a core some 500 kilometers in radius surrounded by a mantle 1,200 kilometers deep and a crust spanning some 60 kilometers. Earth measures 12,756 kilometers across and the Moon spans 3,476 kilometers. The Moon orbits Earth much like the planets orbit the Sun, in an elliptical path. In its monthly orbit, the Moon comes as close as 363,300 kilometers and is as distant as 405,500 kilometers. Because of the physical geometry of the Moon, Earth, and Sun, we see a progression of phases of the Moon as it orbits about us. The lunar cycle begins at New (dark), moves through crescent phases, to First Quarter (one half lit), waxing gibbous, Full Moon (fully lit), waning gibbous, Last Quarter (one half lit), crescent, and back to New. We've already mentioned the coincidence that the Sun and Moon are nearly the same size in the sky, which allows spectacular eclipses. By another peculiar coincidence — the equal time value of the Moon's rotation period and its period of revolution — the Moon always presents the same face toward Earth. The far side of the Moon has been viewed only by spacecraft.

Exploring the Moon with a Small Telescope

Observing the Moon is one of the easiest and most rewarding activities you can do with a small telescope. It is bright, easy to find, has a large, varied surface, and looks impressive when viewed through practically any telescope — or even binoculars, for that matter. To help you find your way around the Moon, invest in a Moon atlas; one of the best is *Atlas of the Moon* by Antonín Rükl (Kalmbach Publishing Co., Waukesha, WI.) It's accurate, colorful, and easy to use.

To the naked eye the Moon looks as if it's a cream-colored disk spotted with dark areas, the markings that form the "man in the Moon." These are maria, latin for seas, so named because they are lunar lowlands that were mistaken for actual seas by early observers. Including Mare Fecunditatis, Mare Crisium, Mare Tranquilitatis, Mare Serenitatis, Mare Imbrium, Mare Nubium,

AT FIRST QUARTER, sunlight illuminates the Moon such that numerous features stand out well for observation. Photo by Alan Dyer.

Mare Humorum, and Mare Frigoris, these areas are indeed lowlands. But instead of beings seas, they are formations in which lava oozed in and cooled in large basins.

The most striking features of the Moon in a small telescope are a number of prominent craters. Perhaps the finest is Copernicus, located northwest of the center of the Moon's disk and conspicuous for its prominent system of rays. As you look at a crater like Copernicus, note how the sunlight illuminates the central mountain peak inside the crater. Another crater with a bright system of rays is Tycho, in the south-central part of the disk. South of Tycho lies Clavius, a large, eroded crater measuring 230 kilometers across. You'll also note a prominent pair of craters near the center of the Moon's disk. These are Hipparchus and Ptolemaeus. East of this pair lies a chain of three prominent craters called Theophilus, Cyrillus, and Catharina. These are but a few of the best craters on the Moon, of which the smallest telescopes will show hundreds.

As you observe the Moon over the cycle of a month, return to some old friends you've looked at before and see how they look now. You'll be amazed at how the view of craters, plains, and other features changes as the angle of sunlight changes. Some of the features look completely different.

The Moon is a steady, reliable companion in the sky. It's there every month waiting to be observed, yet it undergoes occasional special events. When Earth passes directly between the Moon and Sun, we see a lunar eclipse — Earth's shadow cast onto the disk of the Moon. Lunar eclipses can be spectacular

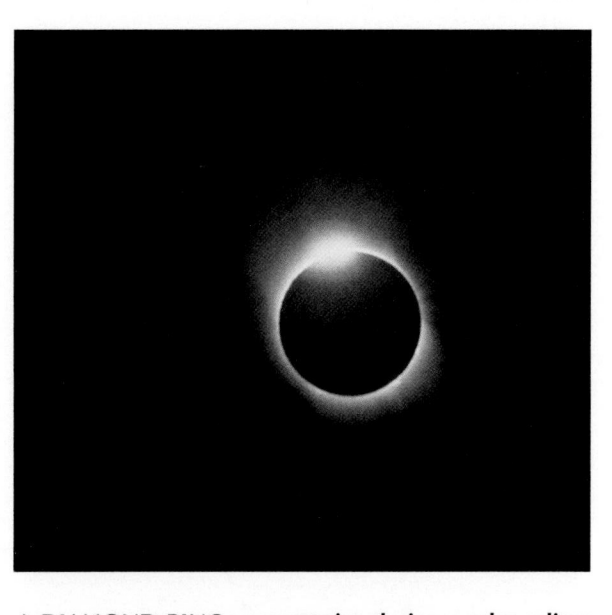

A DIAMOND RING occurs twice during a solar eclipse as sunlight flashes through valleys on the edge of the lunar limb. It is one of the most dramatic moments in astronomical observing. Photo by Jay M. Pasachoff.

because the Moon appears to fade out and often dramatically reddens in color. Partial lunar eclipses occur when Earth's shadow covers only a part of the Moon. Total lunar eclipses — the most dramatic kind — happen when Earth's shadow covers the entire Moon — a "direct hit."

Because of their great brilliance and importance in our lives, the Sun and Moon are central to anyone's observing programs. Check them out occasionally, and remember to observe the Sun only under safe conditions with a proper filter.

5

Solar System Wonders:
Mars, Jupiter, and Saturn

Of the eight planets in the solar system visible in telescopes, three stand out as spectacular. More than their brethren, Mars, Jupiter, and Saturn offer unparalleled detail to observers with small scopes. Particularly near times of opposition every two years, the disk of Mars grows to a relatively large size. This enables backyard astronomers to see the planet's bright white polar ice caps and a number of dark features on the Red Planet's surface. Always relatively large in the eyepiece, Jupiter offers detail in its subtle banding, Great Red Spot, and ceaseless show of orbiting moons. Saturn is perhaps the most breathtaking sight of all in backyard telescopes. Its sharply defined rings inspire virtually everyone who sees it and is the reason many enter the hobby.

Each of these wonders of our solar system is a "must" object — something you must definitely look at and enjoy when they are visible. Yet beyond their claim to startling beauty, Mars, Jupiter, and Saturn have little in common. They are different worlds with different roles in the composition of the solar system.

LARGEST OF THE PLANETS, Jupiter is easily visible in small telescopes, displaying a system of bands, belts, and the typically pale Great Red Spot. Photo by Donald C. Parker.

The Red Planet

The fourth planet from the Sun, Mars measures 6,787 kilometers across, about half the diameter of Earth. The Martian "year" lasts 687 days, during which it orbits the Sun at a distance of between 1.4 and 1.7 astronomical units. Long a favorite of romantic, mystical, and speculative stories, Mars does not harbor intelligent beings or even remnants of canals as proposed by the nineteenth-century astronomer Percival Lowell. The Red Planet, so called for its ruddy color, does have one of the most unusual histories in the solar system.

Mars is a terrestrial planet with a composition much like Earth's. It has a nickel-iron core surrounded by a rocky mantle and a thin crust that we see on the surface. The thin atmosphere is very different from Earth's, consisting chiefly of carbon dioxide with small amounts of nitrogen, argon, oxygen, water vapor, carbon monoxide, and so on. Martian temperatures typically stay below the 0° C mark during the afternoons and at night plummet to -100° C. Although Mars wouldn't be very hospitable to human visitors, it is by far the most Earthlike environment of all the other planets.

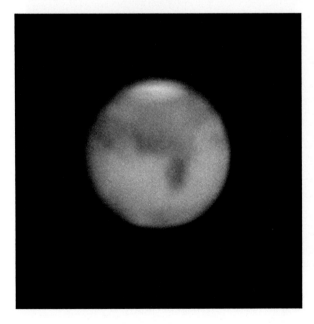

Today Mars is a cold, dry planet, but long ago it was warm and wet. Evidence exists for Martian glaciers, demonstrating that huge amounts of liquid water and snow once existed on the surface. During the so-called Martian Ice Ages, the planet's atmospheric activity was substantially different and may even have supported the conditions for the formation of basic self-replicating organic molecules, the earliest forms of life.

THE RUDDY, DUSTY DISK of Mars shows dark markings — areas of low reflectivity — when viewed with a small telescope. The planet's white polar caps are also visible. Photo by Donald C. Parker.

SATURN'S ENCIRCLING RINGS can be seen in virtually any telescope. Moderate-sized amateur instruments show the prominent dark gap in the rings, called Cassini's division. Photo by Jean Dragesco.

Mars has two small potato-shaped moons, Phobos and Deimos. Discovered in 1877, the moons span 20 by 23 by 28 km (Phobos) and 10 by 12 by 16 km (Deimos). Their odd shapes are easily explained: Phobos and Deimos are asteroids that were long ago captured by Mars' gravity and locked into orbits about the planet.

The visibility of detail on Mars varies considerably depending on where Earth and Mars are in their respective orbits around the Sun. Martian oppositions, times when the planet is opposite the Sun in the sky and is best placed for observing, occur about every two years. Typically Mars measures some 10" or less across in the eyepiece. At times of opposition it may span more than 20" across, making detail on its surface much easier to see. The next great opposition, when Mars will exceed 25" in diameter, will occur in 2003.

The most obvious Martian feature in a small telescope is the planet's northern or southern polar cap (depending on which hemisphere the planet presents to us). The polar caps appear as bright white areas surrounding the poles, and are visible in telescopes as small as 2 inches in aperture. The orange-colored surface of Mars shows many areas that appear dark as well. These surface markings are regions of low reflectivity. Occasionally areas are visible that are slightly brighter than the surrounding terrain, such as when a major dust storm is in progress on the planet's surface.

Prominent areas on Mars include the Hellas and Argyre Basins, impact features that often appear bright due to shifting dust clouds. A huge canyon called Vallis Marineris has been observed on the planet, and its 4,500-kilometer length dwarfs Arizona's Grand Canyon. Nearby lies the Tharsis Ridge, which contains several prominent volcanoes, including Olympus Mons. Although some observers have reported seeing these volcanoes with backyard telescopes,

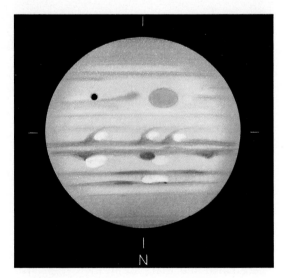

JUPITER'S GALILEAN MOONS, Callisto, Europa, Ganymede, and Io, stand out in any low-power view of the planet. Photo by Ronald E. Royer.

FESTOONS, SPOTS, AND BANDS are visible in ever-changing patterns when viewing Jupiter's cloud features through a backyard scope. Sketch by Jean Dragesco.

TELESCOPES SHOW MARS as a disk covered with light and dark areas, but no canals or signs of life. Even when Mars is distant and its disk small, the polar caps are easy to see.
Sketch by Edwin Faughn.

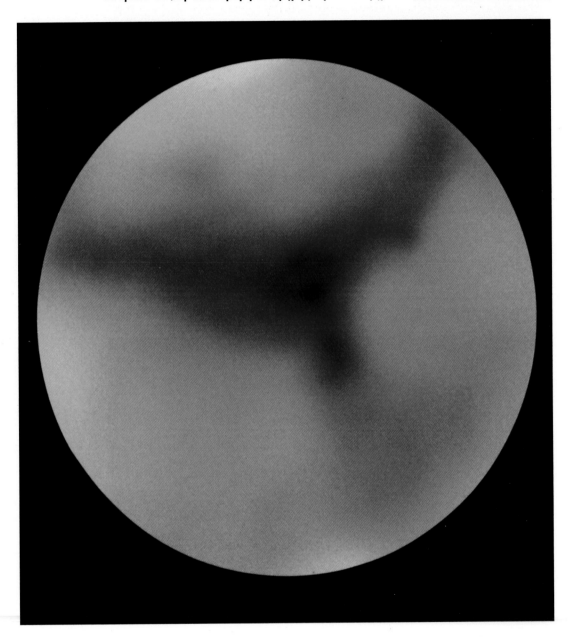

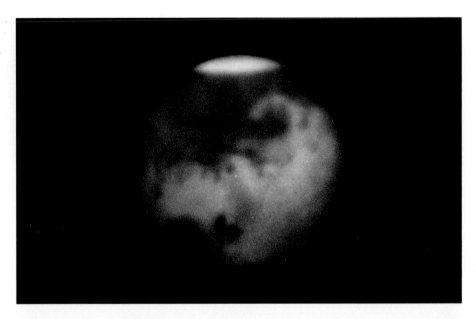

MARS AT HIGH RESOLUTION is a changed world, peppered with small and rapidly changing areas of light and dark. Some of these features correspond to dust storms that sweep the face of the planet. Photo by Jean Dragesco.

they are notoriously difficult to spot. Numerous valleys, ridges, and craters dot the surface. ASTRONOMY magazine in particular regularly provides observing information on Mars, including detailed maps and photographs during each Martian opposition.

King of the Planets

The fifth and largest planet in the solar system is Jupiter, which in essence is a failed star. The prototypical gas giant, Jupiter is an enormous sphere of hydrogen gas that, had it been substantially more massive, may have ignited and become a small star. Had that happened our solar system would orbit a double star system and the orbital dynamics, temperature, and chemistry of the planets would be markedly different. Life on Earth may not have been possible. Fortunately for us, Jupiter lacked the amount of material — by a

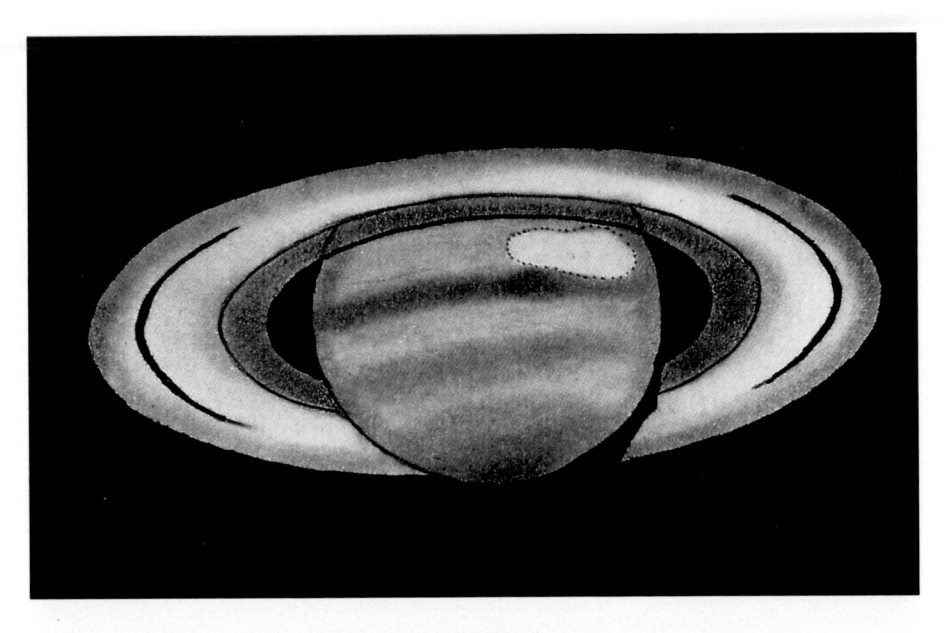

A GREAT WHITE SPOT appeared on Saturn's disk in 1990, and for several weeks observers watched it rotate around the planet. Sketch by Jean Dragesco.

long shot — needed to become a star.

Jupiter measures 142,980 kilometers across at its equator. More than eleven Earths could be stacked end to end across this length. Jupiter is about 1/10 the diameter of the Sun and orbits at an average distance of 5 astronomical units. The Jovian "year" is 11.86 Earth years long. Jupiter is an oblate spheroid. Because of its rapid rotation period of less than 10 hours, the planet's shape is noticeably deformed from that of a perfect sphere.

As with the other gas giants, Jupiter is substantially different in composition from the Earthlike planets. Its extremely low density, 1.3 times that of water, suggests that Jupiter is composed almost entirely of hydrogen and helium. If we could go inside Jupiter we would find incredibly high pressures and temperatures. In fact Jupiter is a heat source — it gives off twice as much heat as it receives from the Sun.

The Jovian "surface" we see in a telescope consists of many layers of thick clouds in the planet's atmosphere. Some 1,000 kilometers below the upper cloud deck lies a sphere of liquid hydrogen, and some 30,000 kilometers

SATURN WITH EDGE-ON RINGS appears much like a smaller version of Jupiter. The occurrence happens when Saturn reaches a point where its rings are exactly edge-on with respect to Earth's line of sight, making the rings disappear. Photo by Charles F. Capen.

below the clouds lies a core made of metallic hydrogen. A rocky core some 10 to 20 times the mass of Earth probably exists at Jupiter's center. The planet has no solid surface like those of the terrestrial planets.

Coated by thick layers of clouds and driven by internal heat, rapid rotation, and high pressure, the Jovian atmosphere is one of the most complex known. Reddish-brown belts and zones cross the planet parallel to the equator. These have gas flows that descend far below. White and yellowish zones are high-altitude cloud decks. Weather systems form and stay in zones, creating the many bands and belts visible on the planet. The equatorial region of Jupiter contains enormous anticyclonic (clockwise) storms like the Great Red Spot, a storm larger than Earth that has existed for hundreds of years. High clouds contain vast amounts of ammonia while lower clouds consist of sulphur, hydrogen, and phosphorous compounds.

Sixteen moons orbit the planet like a solar system in miniature. They are Metis, Adrastea, Amalthea, Thebe, Io, Europa, Ganymede, Callisto, Leda, Himalia, Lysithea, Elara, Ananke, Carme, Pasiphae, and Sinope. Of these, four so-called Galilean moons are by far the brightest and are immediately notice-able in backyard telescopes. Discovered by Galileo in 1610, Io, Europa, Ganymede, and Callisto are so large that big amateur telescopes on nights of steady seeing can show these moons not as points of light but as tiny disks. The diameters of these moons are 3,630 kilometers (Io), 3,050 km (Europa), 5,260 km (Ganymede), and 4,800 km (Callisto). The Galilean moons are among the largest in the solar system. Ganymede is, in fact, larger than Mercury and Pluto, and all four of the Galilean moons would be considered planets if they orbited the Sun rather than Jupiter. At an orbital distance of 55,000 km, the Jovian sys-tem has another feature, a set of rings. Discovered by the Voyager 1 spacecraft in 1977, the rings are composed of rocky boulders and represent a moon that broke apart long ago. Because it is so thin, this ring is not visible from Earth.

In amateur telescopes Jupiter is a delight. The first thing you'll notice are the bands and belts, some appearing darker and in stronger colors than others. If the Great Red Spot is facing Earth you may see it too, appearing as an oval with a pale red or pink color. The planet's overall color is typically described as ocher, but it may appear somewhat yellowish-brown. Also, take note of the moons — especially the four bright Galilean moons. You may see them eclipsed by Jupiter or pass in front of the planet, casting a shadow onto the upper cloud deck, as they slowly revolve.

The Ringed World

Although it doesn't offer telescopists as much cloudtop detail as Jupiter does, Saturn may be even more inspiring than its fellow gas giant. Saturn and its rings comprise one of the greatest sights in astronomy, one that will stick with you long after you've first seen it. The sixth planet from the Sun, Saturn measures 120,540 kilometers in diameter, over nine times Earth's diameter and

AN ELECTRONIC SATURN is a colorful world composed of numerous pixels stored in a computer. This CCD image of the planet was made with a 16-inch telescope and software designed for processing astronomical images. Photo by Donald C. Parker; image processing by Richard Berry.

1/11th the diameter of the Sun. Saturn orbits the Sun at a distance of 9.5 astronomical units and completes one Saturnian "year" every 29.46 years. Saturn's famous rings, the planet's most obvious feature, span 600,000 kilometers — nearly half the diameter of the Sun.

One of the most curious aspects of Saturn is its low density, which is less than that of water. If one were to place Saturn in a giant bathtub filled with water, it would float. Like Jupiter, it is composed chiefly of hydrogen and helium with smaller amounts of methane, ethane, ammonia, and water. It probably has an iron-laden rocky core and an outer core of ammonia, methane, and water. This is encased in liquid metallic hydrogen which in turn lies deep within liquid molecules of hydrogen. Like Jupiter, Saturn has no solid surface to speak of and when we see it in a telescope we are looking at the planet's cloud-tops.

Also like Jupiter, Saturn displays a range of weather phenomena in its layers of clouds. Saturn radiates heat from within and this drives storms, belts, and zones of clouds like those on Jupiter. Saturn's atmospheric activity is far more subtle, however, and not as easily observed with earthbound telescopes.

In addition to the rings, Saturn has 18 major moons: Pan, Atlas, Prometheus, Pandora, Janus, Epimetheus, Mimas, Enceladus, Tethys, Telesto, Calypso, Dione, Helene, Rhea, Titan, Hyperion, Iapetus, and Phoebe. By far the largest and brightest of these is Titan, which is visible as a tiny "star" near the planet on any good night of observing. Titan measures 5,150 kilometers across, making it one of the largest moons in the solar system and larger than Mercury and Pluto.

When you observe Saturn, take note not only of the rings but of subtle details on the globe of the planet itself. Occasionally white spots and subtle bandings are visible that indicate storms and other weather activity in the cloudtops.

Saturn, Jupiter, and Mars will give you hours of enjoyment when viewed with your telescope. They are the greatest wonders of the solar system, filled with great sights and changing detail.

6

The Inner Planets: Mercury and Venus

In an approximate sense the solar system has three main constituents: a set of inner, rocky bodies, a group of large gaseous planets, and an outer cloud of planets, miniplanets, and cometary nuclei made up almost entirely of dirty ice. Of the rocky or terrestrial planets close to the Sun, we should really include Mercury, Venus, Earth, and Mars. Because we needn't discuss observing Earth with a telescope, and because we examined Mars in the previous chapter, we'll now focus attention on the two closest planets to the Sun, Mercury and Venus.

Although these two are relatively close to Earth, before 1965 astronomers didn't know a great deal about them. The smallest features known on Mercury and Venus were several hundred kilometers across. But the advent of spacecraft missions to the inner planets beginning in 1965 painted a detailed picture of these worlds as intensely hot, inhospitable places with marked contrasts.

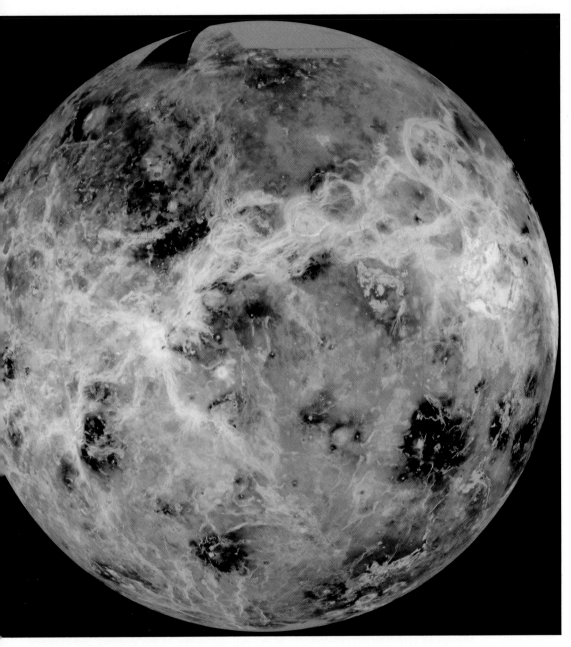

VENUS IN FALSE COLOR RADAR appears suitably hostile: the planet is hellishly hot and cloaked in a thick atmosphere of carbon dioxide. NASA image.

VENUS IS THE BRIGHTEST "STAR" in the sky, appearing in the evening and morning when it outshines everything except the Moon. In this Paris evening sky, Venus lies above and left of the Moon. Photo by Olivier de Goursac.

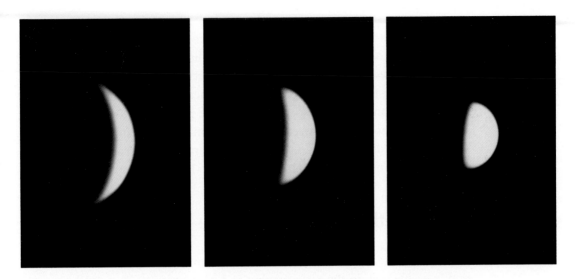

THE PHASES OF VENUS are observable in a small telescope. They occur because the planet lies between Earth and the Sun, and the changing orbital geometry shows you an entire range of partial phases. Photos taken over a four-month span by Jean Dragesco.

The Iron Planet

Mercury is the closest planet to the Sun. It measures 4,878 kilometers across — less than half Earth's diameter — and orbits the Sun at a distance of about 0.4 astronomical units. It is less than half as far to the Sun as Earth. Mercury's "year" lasts 88 days. The planet has no moons. Imagery from spacecraft show that Mercury resembles the Moon more than any other solar system body. It is cratered, barren, and without much variety of geological features. If Mercury had an early pre-atmosphere, it was vaporized long ago by impact bombardment and by the Sun's radiation stream. It is a stark world located in an inhospitable place.

More than any other body in the solar system, Mercury deserves recognition as the iron planet. It consists of a large nickel-iron core that probably measures 3,600 kilometers across. The planet's mantle and crust, then, are together possibly only 600 kilometers thick. Although much of Mercury's surface resembles the Moon, unlike the Moon, it contains large, open plains in its highlands. Because of Mercury's higher gravity, it contains fewer large craters than the Moon because the impact debris is constrained into a smaller area. Mercury also contains a large number of peculiar cliffs that span hundreds of kilometers. These may have formed as the planet's red-hot iron core cooled and shrank, pulling the planet's "skin" closer together. The most notable single feature on the planet is the Caloris Basin, a huge impact crater stretching 1,400 kilometers across. Because Mercury lacks an atmosphere to moderate temperatures, sunlight directs the wildly variable temperatures. Afternoon highs usually hit -200° C; night temperatures plummet to -180° C.

Viewing Mercury with Eye and Telescope

Because Mercury lies so close to the Sun it is fairly difficult to observe, and is visible only in the early evening and early morning skies. The planet is relatively bright, typically shining at about magnitude 0, making it visible to the naked eye. Because it strays only a short distance from the Sun, you'll most often see it not in total darkness but while the sky is in twilight after sunset or before sunrise. A pair of binoculars is handy for initially spotting Mercury's soft, orange glow.

Telescopically, it is a challenging object. The lack of contrast in a twilight sky and the planet's small diameter (typically between 5" and 10") makes seeing detail on Mercury's surface difficult. Yet experienced planetary observers using long-focal-length telescopes occasionally report shadings and areas of subtle bright patches on the planet's disk. Because of the orbital geometry of Sun, Mercury, and Earth, we see Mercury go through phases as we and it orbit the Sun. The chief observational challenge with Mercury is to see it as its phase changes, and you may wish to sketch or photograph the planet in a variety of phases.

Planet of Fire

Second in closeness to the Sun, Venus is a substantially different place than its neighbor Mercury. Venus has a thick, dense atmosphere composed mostly of carbon dioxide. Early thoughts that Venus was Earth's "sister planet"

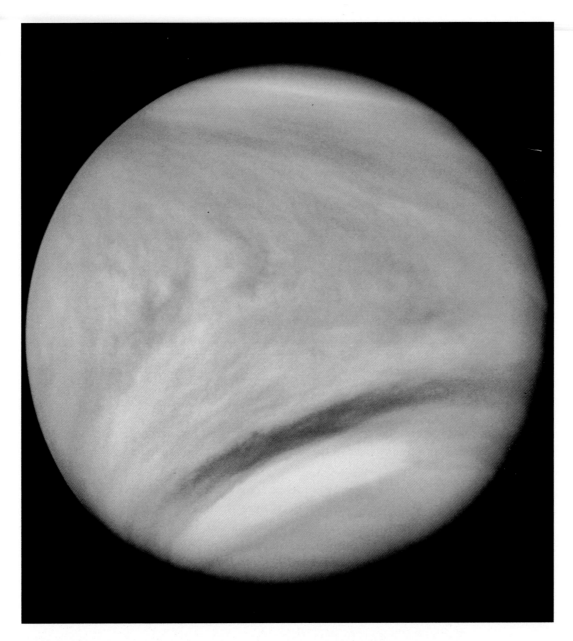

THE THICK CLOUDS OF VENUS make seeing surface detail on the planet impossible, even from spacecraft. Radar is the only means to penetrate the cloak of haze. NASA photo.

VENUS IS SO BRIGHT it is possible to see and photograph it in the daytime. This photo was made in the late afternoon of May 29, 1988. Photo by Mark Payne-Gill.

collapsed as spacecraft analyzed it over the past two decades. Instead of a world similar in many ways to Earth, astronomers found the most hellish place in the solar system.

Venus is 12,104 kilometers across, just shy of Earth's diameter. Its density is lower than Earth's; thus, its mass is smaller. Venus orbits the Sun at an average distance of 0.7 astronomical units. The Venusian "year" lasts 225 days.

Strangely, the internal composition of Venus may be similar to Earth's. However, in two critical places — surface and atmosphere — Venus is radically different. The prototypical example of the Greenhouse Effect is at work on Venus. Not only is the planet close enough to the Sun to be bombarded by an intense stream of radiation and heat, but its thick atmosphere acts as a blanket that traps the heat, further raising the temperature and making possible complex chemistry high in the Venusian clouds, chemistry that among other things produces voluminous toxic gases like sulfur dioxide.

MERCURY'S BATTERED SURFACE strongly resembles that of the Moon, though Mercury is even more heavily cratered. NASA photo.

THE MOON AND MERCURY gracefully dance above the horizon in a photo made on January 9, 1989, near Flagstaff, Arizona. Photo by Brian A. Skiff.

As if that weren't offputting enough for living beings, the temperatures on Venus are extreme. The average temperature at the top of the 250-kilometer high atmosphere is 26.8° C, and the temperature down on the surface is some 476.8° C, two-and-one-half times those on Earth. To make matters worse, Venus' surface pressure is a crushing 90 times that of Earth's. It would be very difficult to visit the surface of Venus, let alone inhabit it. Again, we see evidence of how lucky we are to have Earth!

Unlike Mercury, Venus displays a fabulous range of geological structures. Large rolling plains blanket about 70 percent of the planet's surface, lowlands cover some 20 percent, and highlands occupy 10 percent of the topography. Many craters have been identified, the result of impacts that occurred before Venus' atmosphere developed to its present density. Two immense areas of volcanic activity exist on the planet, Ishtar Terra and Aphrodite Terra. The highest mountain is Maxwell Montes, which rises 10 kilometers above the planet's surface.

TINY MERCURY TRANSITS ACROSS the face of the Sun in a rare occurrence. Despite the Sun's large disk, Mercury infrequently places itself precisely between the Sun and Earth. Photo by Jean Dragesco.

The Changing Phases of Venus

Like Mercury, Venus never strays far from the Sun in our sky. It is far brighter than Mercury, however — in fact Venus is the brightest object in the sky after the Sun and Moon — and therefore it is easily recognized as the "evening star" or "morning star" at twilight. At a peak magnitude of about -4.5, Venus is so bright that it often confuses experienced scientists, or pilots, or other professionals, suggesting the appearance of a "UFO." Venus is so bright because it is close to Earth and because its opaque clouds reflect sunlight extremely efficiently.

Also like Mercury, it displays phases. Although Venus is much closer and appears substantially larger in the sky than Mercury — up to 60" across — the Venusian clouds make spotting details on the planet virtually impossible.

As with Mercury, skilled planetary observers sometimes report seeing subtle shadings and differences in brightness on the planet's disk. But as with Mercury the challenge here comes in viewing Venus over a great range of phases, from Full, when the planet appears small, to a slender crescent, when it is large.

Though they do not offer bands, rings, belts, or other surface detail like the spectacular planets Mars, Jupiter, and Saturn, Mercury and Venus are well worth observing. They are morning and evening friends whose calm, steady appearance in the telescope hides a barren, stark world on Mercury and a world of inhospitable heat and gas on Venus.

VENUS AND MERCURY in telescopes sometimes appear like small, bright crescents, as Venus did in this photo taken on June 2, 1988. Photo by Martin C. Germano.

MERCURY PEEKS ABOVE THE HORIZON in a crystal clear sky on September 13, 1978. The planet's frequent low altitude above the horizon makes it tricky to observe. Photo by Leo Enright.

7

The Outer Planets:
Uranus, Neptune, and Pluto

In the middle of the eighteenth century, most astronomers thought knowledge of the solar system was pretty complete. Six planets inhabited the solar system, slowly orbiting the powerful central Sun. Then in 1781 a German-born, English amateur astronomer named William Herschel discovered a seventh planet and in one night turned orderly knowledge into chaos. The planet Herschel discovered was Uranus, and it lay more than twice as far from the Sun as Saturn, the previous outermost planet. This doubled the size of the known solar system. The business of astronomy called outer solar system research was born, and it continues today. In fact, in 1992 astronomers discovered a 200-kilometer-sized object more distant than Pluto. Provisionally designated 1992

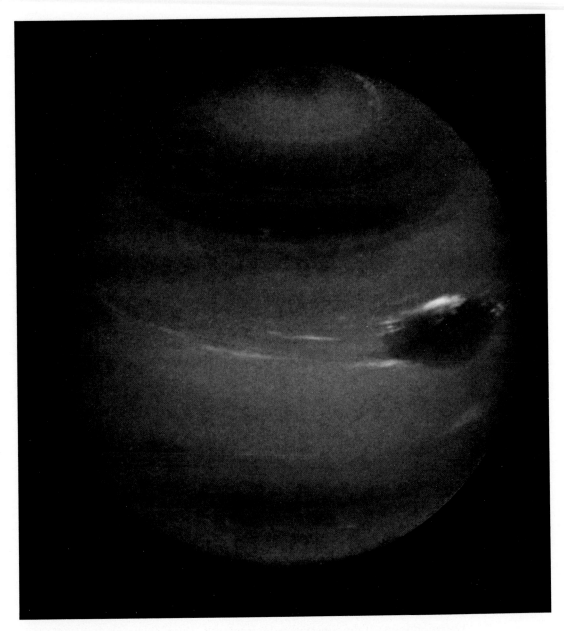

BIG, BLUE NEPTUNE displayed an amazing array of Earthlike weather phenomena to the Voyager 2 spacecraft in 1989, including a prominent dark oval reminiscent of Earthbound weather and cirrus-like clouds. NASA photo.

NEPTUNE AND MOON TRITON appeared like blue-green and golden disks as the Voyager 2 spacecraft approached them in the spring of 1988. NASA photo.

QB$_1$, the object may be a "miniplanet" that represents a new class of distant solar system bodies.

In 1846 J.G. Galle at the Berlin Observatory discovered the eighth planet, Neptune. This planet, astronomers found, is half again as far out as Uranus. And in the ongoing search to find even more distant planets, Clyde Tombaugh in 1930 found the ninth planet, Pluto, at Lowell Observatory in Arizona.

The discoveries of Neptune and Pluto relied on complex mathematical calculations and predictions of where such planets might be found. Typically Uranus glows slightly brighter than magnitude 6, making it just visible to the naked eye on a superbly dark night. Yet the multitude of 6th-magnitude objects hid the planet from discovery until 1781. Neptune glows at 8th magnitude, making it visible in binoculars but invisible to the naked eye. Pluto is extremely faint at magnitude 14, making it barely visible in a good 8-inch telescope.

Herschel's World

The discovery of Uranus revolutionized astronomers' ideas of the solar system, because it demonstrated the immense size of the Sun's family of planets. Uranus lies at an average distance from the Sun of 19 astronomical units — nearly four times more distant than Jupiter. Like Jupiter and Saturn, it is a gas

A TELESCOPIC VIEW OF NEPTUNE shows a small, round disk glowing at about 8th magnitude. The planet's blue-green color helps betray its presence in small scopes. Photo by Geoff Chester.

giant planet composed primarily of hydrogen gas in a giant sphere. The diameter of Uranus is 52,120 kilometers, more than four times that of Earth. Uranus has a "year," one Uranian orbit about the Sun, that lasts 84 years. Curiously, it is tilted 98° with respect to its orbital plane, meaning that at various times during its orbit it displays its poles — not its equator — toward the Sun. It is a planet that has been "knocked on its side."

Uranus has a rocky core some 16,000 kilometers in diameter surrounded by an 8,000-kilometer thick ice mantle. Surrounding this is the part we see, a layer of molecular hydrogen enveloped by a thick atmosphere of hydrogen clouds. This is typical of a gas giant. Uranus also has a thin set of rings and fifteen moons, Cordelia, Ophelia, Bianca, Cressida, Desdemona, Juliet, Portia, Rosalind, Belinda, Puck, Miranda, Ariel, Umbriel, Titania, and Oberon. Of these, Miranda is the most fascinating. It displays so many different types of planetary surfaces in a single, small moon that astronomers believe it is the result of collisions of large bodies and subsequent activity by impact and flowing surface material.

In a telescope Uranus appears like a small, bluish disk about 3.5" across. At low power you may recognize it best by the color, and you may have to switch to a moderate magnification to see the planet's disk. Although cloud-top details are not visible in small telescopes, Uranus offers a thrill simply because it is so distant.

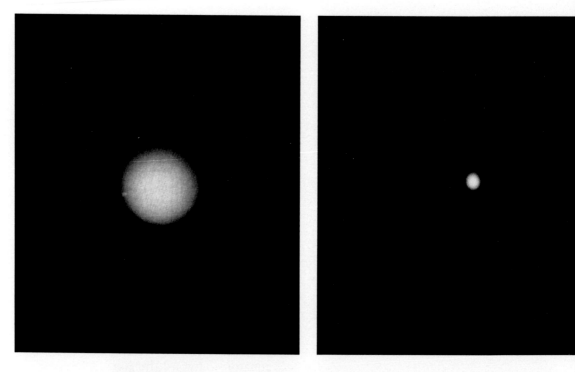

GREENISH URANUS appeared like a featureless ball to the Voyager 2 spacecraft in 1986. NASA photo.

URANUS IN A TELESCOPE appears somewhat larger and brighter than Neptune, and is chiefly identifiable by its strong color. The planet is barely visible to the naked eye on dark nights. Photo by Geoff Chester.

Surprising, Dynamic Neptune

Neptune, the nineteenth century's contribution to new planets, is on a gross scale quite similar to Uranus. It is a large, blue-green gas giant with a predominantly hydrogen composition. The planet spans 49,530 kilometers across, slightly smaller than Uranus. Neptune's "year" is 165 years long. Because it is so much more distant — it lies an average distance of 30 astronomical units — Neptune is correspondingly smaller and fainter as we see it in telescopes. The composition is similar to that of Uranus, with a rocky core, icy mantle, and dense, thick atmosphere composed of many cloud decks. The Voyager 2 space-

Uranus

URANUS AND SATURN passed close to each other in the sky during a rare conjunction in 1988. The Lagoon and Trifid nebulae occupy the central part of the image; Saturn is the overexposed "star" at top, while Uranus is marked with an arrow. Photo by Jean Dragesco.

NEPTUNE'S MOON TRITON shows an uneven, "cantaloupe" terrain. NASA photo.

craft showed a great surprise at Neptune that it failed to show at Uranus: Neptune displays an incredible variety of weather phenomena. Some of these, such as large cyclone-like storms and groups of high-altitude cirrus, appear almost Earthlike. The planet's atmospheric temperature, however, is decidedly chillier than Earth's: -220° C.

Neptune has eight moons: Thalassa, Naiad, Galatea, Despina, Larissa, Proteus, Triton, and Nereid. The largest of these by far is Triton, which is bright enough to see in backyard telescopes. The planet itself is slightly brighter than 8th magnitude, making it visible in binoculars and finder telescopes. Like Uranus, it is chiefly identified by its blue-green color (Neptune appears more green than does Uranus.) In telescopes Neptune shows a 2.3" disk on which no detail is visible. Again, with the outer planets, the challenge is seeing something that is so distant.

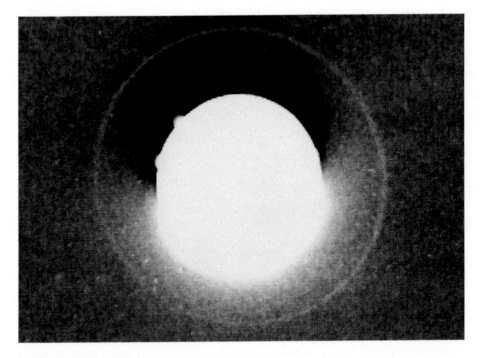

ENCIRCLING URANUS (center, overexposed) is a band of material called the Epsilon Ring, which was first clearly shown by the Voyager 2 spacecraft in 1985. NASA photo.

The Frozen Double Planet

Few planets have remained as mysterious as Pluto, the outermost and most recently discovered. All the planets have been visited by spacecraft except Pluto. In addition, its composition — as much as can be discerned from Earth-based telescopes — is markedly different from the other outer solar system planets. Rather than a gas giant, Pluto is a small, icy world. Some astronomers believe that long ago the outer solar system was filled with thousands of Pluto-like bodies, relatively small, icy chunks of matter. Similar objects include Neptune's moon Triton and 1992 QB$_1$. Between the ancient history of the solar system and today, astronomers believe, the outer solar system was swept mostly clean of these objects and they were pushed out well beyond the orbit of Pluto.

Pluto orbits at an average distance of 39 astronomical units, but

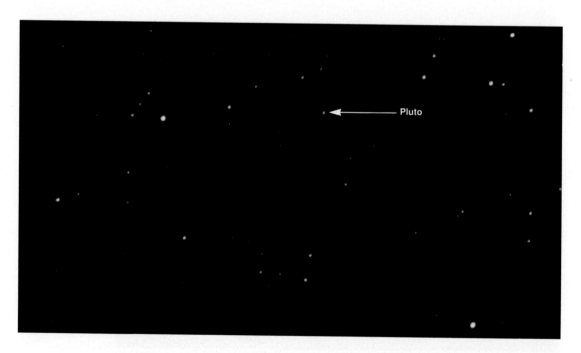

PLUTO IS A FAINT "STAR" in backyard telescopes (arrow). It glows dimly at nearly 14th magnitude and a 6-inch scope is the minimum required to see it. Photo by Dale K. Bonges.

PLUTO AND MOON CHARON are barely separated in images made with the best telescopes on Earth. Their secrets will remain until a spacecraft arrives at the distant system. Photo courtesy Canada-France-Hawaii Telescope Corp.

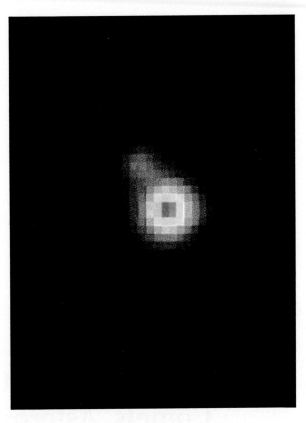

because its 248-year-long orbit is highly elliptical, it is not always the most distant planet. During portions of its orbit it is actually closer to the Sun than Neptune. The diameter of Pluto is not well established, but is probably about 2,300 kilometers — about half the size of Mercury. In 1978 James W. Christy at the U.S. Naval Observatory discovered a tiny object orbiting Pluto and named it Charon. This small moon orbits at a distance of 20,000 kilometers and may be half Pluto's size, making the Pluto-Charon pair the most equal in size of any planet and moon.

Because its surface temperature is about -230° C, Pluto is too cold to have much of an atmosphere. But during times when it is somewhat closer to the Sun than average, astronomers believe the planet may warm enough to release gases into a temporary atmosphere that surrounds both Pluto and Charon. The makeup of Charon is similar to that of Pluto — dirty ice.

Pluto is so small and distant that it is barely observable with backyard telescopes. At its brightest the planet appears like a faint star of about 14th magnitude, requiring a finder chart to pick it out from the surrounding stars in the field of view. Consult a star chart in one of the amateur astronomy magazines to discover how to find Pluto for that year. Finding and identifying it by its slow motion relative to the stars is the ultimate in planet hunting. When you see Pluto in the eyepiece, you'll appreciate why the planet remained undiscovered until this century.

8

Wanderers of the Solar System: Comets, Asteroids, and Meteors

So far we've surveyed virtually all of the solar system. The Sun, inner planets Mercury and Venus, the Earth-Moon system, showpieces Mars, Jupiter, and Saturn, and the outer planets Uranus, Neptune, and Pluto. These major planets along with their moons and our Sun constitute the main pieces of the solar system. But other, smaller chunks of rock and ice float about the Sun in a variety of orbits and round out our understanding of how the solar system works.

The micromatter in our solar system shows up in a variety of different species. Comets are chunks of dirty ice that live deep in the outer solar system. Although most comets exist far beyond our ability to see them and stay in a permanent deep freeze, some orbit the Sun closely enough that they heat up on

COMET WEST blazed across the sky in 1975-1976 and was easily visible to the naked eye. On the average, such bright comets appear once a decade. Photo by Ronald E. Royer and Steven Padilla.

occasion and produce glowing tails. Asteroids, or minor planets, are chunks of rock that orbit far from the Sun — many in a zone between Mars and Jupiter — and range in size from 1,000 kilometers across to the size of a refrigerator. Asteroids usually wander harmlessly as they circle the Sun, yet an asteroid impact on Earth 65 million years ago likely killed many life forms including the dinosaurs, and at some date an asteroid will hit Earth again.

The smallest particles in the solar system can be seen in Earth's atmosphere as they enter, heat up, and vaporize. Watching such meteors — "shooting stars" — makes a pleasant evening activity. Most of these are no larger than a grain of sand. Large particles that strike Earth's surface are called meteorites. Other phenomena related to small solar system particles can also be observed. The zodiacal light is a subtle brightening visible along the ecliptic, the apparent path of the planets across the sky. It is caused by sunlight scattering off small particles in the plane of the solar system. Even harder to spot is the gegenschein, or counterglow, a faint spot of light caused by particle scattering directly opposite the Sun's position in the sky. Finally, energetic particles from the Sun itself enter Earth's magnetic field, spiral inward and speed up, creating a ghostly glow called an aurora. The effect is best visible after a solar storm and from a spot near Earth's North Magnetic Pole (aurora borealis, or northern lights) or South Pole (aurora australis, or southern lights).

Icy Mudballs in the Frozen Deep

Comets are one of the most awe-inspiring sights in nature. Taken by early skywatchers as portents of doom or impending change, comets are in fact frozen garbage left over from the accretion of the major planets. A chunk of frozen mud and ice, the nucleus of a comet — the permanent, solid part — probably measures a few kilometers across, the size of a small city. Comets contain two other distinct parts, particularly when they pass relatively close to the Sun and warm up. These are a coma, an extended halo of particles vaporizing from the nucleus, and a tail, a stream of dust and ionized gas from the nucleus that always points away from the Sun. The Sun's radiation and its solar "wind" of energetic particles push away material from the cometary nucleus. A comet's tail can stretch for millions of kilometers, although its density is low.

A BRILLIANT METEOR FLASHES past the Lagoon Nebula in Sagittarius, leaving a streak on a time exposure of the constellation. Photo by Leonard C. Pincus.

COMET BRADFIELD, visible throughout much of 1987, showed in addition to its tail a distinct antitail, a sunward spike. Photo by Chris Schur.

HALLEY'S COMET, though not as bright in 1986 as the public had hoped, was nonetheless a naked-eye comet under dark skies. Photo by Gordon Garradd.

Comets orbit the Sun in two broad classes, long-period and short-period. Short-period comets have orbital periods of 150 years or less. Most of them lie within the orbits of the nine planets and, indeed, were drawn into the solar system by the gravitational effects of the planets. Long-period comets have elliptical orbits up to millions of years — and some of them come in close to the Sun and are catapulted away never to return. Astronomers believe that most comets exist in the Oort Cloud, a spherical stockpile of cometary nuclei far beyond the orbit of Pluto. (The feature was proposed by the Dutch astronomer Jan Oort.) This is the residence of long-period comets, which occasionally through the tug of gravity get kicked in toward the Sun. Short-period comets may reside in a region called the Kuiper Disk, a dense, flat cloud of debris not too far beyond the orbit of Pluto.

The most famous of all comets is Halley's Comet, which last came in close to the Sun in 1985/1986. Unfortunately the orbital geometry of Earth, comet, and Sun was arranged so that the comet did not become as bright as it has on many previous apparitions. Halley's 76-year period will bring it into view again in the year 2061, when it will be better visible. Another notable comet, Swift-Tuttle, was recovered in 1992 after 130 years. This comet is the parent of the famous Perseid meteor shower which is visible every August. The last spectacular comet, Comet West, appeared in 1975/1976, and was prominent in the morning sky. Shining at magnitude 0 and with a tail that stretched across many degrees on the sky, it was a striking sight to the unaided eye. If the laws of probability pay off, we're due for another bright comet at any time.

Several times during each year telescopic comets are visible. They may not be impressive naked-eye sights, but many comets in the 8th-10th magnitude range show a fuzzy coma and a short tail when viewed with even a 3-inch telescope at low power. Consult the latest astronomy magazines to see if any comets are currently visible in the night sky.

The Orbiting Rocks

Asteroids are also solar system debris, but they differ from comets in fundamental ways. They are rocky and probably 95 percent of them orbit in a disk situated between Mars and Jupiter — at distances of 2.2 to 3.3 astronomi-

THE PERSEID METEORS, which peak each year on August 12, are often bright and yellow in color, as is this Perseid that swept past the Double Cluster. Photo by Richard Andreasson.

A BRILLIANT AURORA shimmers over the town of Kiana, Alaska, on February 10, 1982. Such aurorae, often called the Northern lights, are caused by particles from the Sun interacting with Earth's magnetic field. Photo by Forrest C. Baldwin.

cal units — called the main asteroid belt. Like planets, most of these main belt asteroids have orbits that are nearly circular. Astronomers believe that 100,000 asteroids may exist within the main belt, but they have discovered and named only about 5,000 with well-known orbits. In all likelihood asteroids are the debris from a group of miniplanets that once existed between Mars and Jupiter but fragmented in a series of collisions.

The first discovery of an asteroid came on January 1, 1801. On that New Year's Day the Italian astronomer Giuseppe Piazzi found 1 Ceres, which turns out to be the largest asteroid with a diameter of 1,000 kilometers. This discovery opened up a new age of solar system exploration during which many other astronomers searched for and found other asteroids. The most famous of these are the "big four"; Ceres, 2 Pallas (discovered in 1802; 540 kilometers), 3 Juno (1804; 247 km), and 4 Vesta (1807; 515 km). Most asteroids are much smaller than these, however: only 200 exceed 100 kilometers in diameter.

The brightest asteroid, Vesta, is just barely visible to the naked eye. But realistically the brightest few are best visible in small telescopes. The small diameters of asteroids makes them appear starlike in earthbound telescopes. The brightest glow at 6th and 7th magnitude, and many dozens are in the comfortable range of backyard observing at 10th to 12th magnitude. Because they appear stellar, asteroid observing is tricky. You must identify the field of view in which an asteroid lies, make a quick sketch of the field, and come back much later during the night or the following day and resketch the field. The "star" that has moved relative to the background stars is not a star, but the asteroid.

Not all asteroids (and comets, for that matter) have

ASTEROID 44 NYSA (arrow) appears like an ordinary star as it slowly passes through the Beehive Cluster in Cancer. Photo by Lee C. Coombs.

A LONG, KINKY GAS TAIL characterized 1989's Comet Brorsen-Metcalf, visible in a pair of binoculars at the time this photo was taken. Photo by Chris Schur.

AS BRIGHT AS THE FULL MOON, a bolide meteor falls toward the horizon during the early morning hours of November 25, 1989. Photo by Gordon Garradd.

innocent orbits. Over millions and billions of years, many solar system rocks will come close to or intersect the orbits of other asteroids, comets, or even planets. The evidence of impacts in the solar system is abundantly obvious on the Moon, Mercury, and other impact-scarred planets and moons. Earth, indeed, wears its own scars from long-ago impacts, such as the Barringer Meteor Crater in Arizona. Unlike barren worlds, however, Earth's active weather system helps to cover up old scars. The message is clear, however: although much of the debris of the early solar system is gone, impacts will occur in Earth's future. The difference between the future and the past, however, may be that earthlings now have the technology to discover objects on a collision course with Earth and possibly nudge their trajectory, turning a direct hit into a miss.

Raining Pebbles on Earth

Rocky particles much smaller than asteroids float calmly around the Sun in the millions. Sources of these microparticles include asteroid collisions that send small packets of debris into odd orbits and the trails of comets, whose particles stream off the cometary nucleus and accumulate in their orbits. When Earth crosses into the orbit of a comet or when scattered debris falls into Earth, the small particles are pulled in by Earth's gravity and burn up in the atmosphere. As the particles burn up one of the energy byproducts is light, so we see them as meteors, brief flashes of light that streak across the sky.

Meteor watching is one of the most enjoyable aspects of amateur astronomy. It requires nothing but a chaise longue or sleeping bag on which to sit or lie and a steady gaze up at the sky. Random or sporadic meteors are visible on every clear night, but meteors are visible in greater numbers during so-called meteor showers. These occur when Earth's orbit intersects a stream of debris, revealing many meteors per hour.

Twenty principal meteor showers have been identified. During a shower the meteors appear to come from a radiant, a particular point in the sky, but the meteors can streak across in any direction from that point. By far the best time to observe meteors during any shower (or on any random night) is between local midnight and dawn, when Earth's rotation turns it head on into the cloud of debris.

The first meteor shower of the year is the Quadrantid shower (which peaks on January 4), an event that boasts as many as 110 meteors per hour during its very short peak. Other exceptional showers include the April Lyrids (April 22; 12 meteors at peak), the Eta Aquarids (May 5; 20), the Delta Aquarids (July 27; 35), the Perseids (August 12; 70), the Orionids (October 21; 30), the Taurids (November 8; 12), and the Geminids (December 14; 60). Of these, the most popular meteor shower is the Perseid event, which occurs in midsummer and is notable for its many bright meteors that leave glowing trails visible for several seconds. The Perseid shower has been remarkably active during the past few years, and in 1992 Comet Swift-Tuttle, the comet responsible for the Perseid debris, was rediscovered after 130 years.

Almost all meteors that enter Earth's atmosphere burn up long before reaching Earth's surface. Meteors that strike Earth are called meteorites, and thousands exist. Some amateur astronomers have become ardent meteorite collectors, in fact. Meteorites exist in two principal types, stony and nickel-iron.

Strange Lights in the Night Sky

Some solar system particle debris acts on the night sky in more subtle ways. One of the most stunning spectacles in the sky is a bright aurora, a dancing, shimmering display of greenish, yellowish, or even ruddy light that may be visible in the northern sky or spread across the sky's dome. (South of the equator the aurora is visible in the direction of the south pole.) Aurorae can be subtle or startlingly bright. One of the most striking aurorae in recent times was visible on the night of November 8, 1991, when Northern Hemisphere observers saw the entire sky explode in a cinematic upheaval of curtains, pillars, streaks, coronae, rays, and other bright, well-defined and multicolored glows. Aurorae result from bursts of solar particles that enter Earth's magnetic field and interact with atmospheric molecules, giving off photons as they spiral down toward the poles at super high speeds.

Other haunting skyglows are more subtle. The zodiacal light is a diffuse band of milky light that is visible only under a dark sky. Caused by sun-

BRILLIANT SHADES OF RED characterize this aurora surrounding the South Celestial Pole, photographed from Australia on October 20, 1989. Photo by Gordon Garradd.

light scattering off dust particles in the plane of the solar system, the zodiacal light stretches across the ecliptic, the apparent path of the planets on the sky. A similar phenomenon is the gegenschein, or counterglow, a subtle glow on the sky opposite the position of the Sun. You may have to wait for an exceptionally clear, dark sky to see them though.

9

Stars: Single, Double, and Variable

On any clear night we can gaze up at the sky and see thousands of stars, so bright and seemingly so close yet in reality astonishingly distant. Fascination with stars is hardly new, having existed since the first high order mammals looked skyward. Yet only in the 20th century have astronomers made substantial headway in figuring out exactly how they form, live, and die.

Stars are the basic units of matter in the universe. They are energetic balls of mostly hydrogen gas that slowly fuse their material into heavier elements until one day they exhaust the fuel stores and end their lives. The Sun is of course the closest star, the powerful source of energy in our solar system. As we've seen in a previous chapter, the Sun is an average star. Stars exist in many sizes, temperatures, colors, and chemical compositions, each allowing astronomers to piece together a little part of our knowledge about the universe.

STAR COLORS cover the spectrum from blue to white to yellow to orange to pale red. In addition to an array of different colors, stars come in all brightnesses, too. Photo by Jim Barclay.

A GALAXY OF STARS awaits amateur astronomers every night, and so many exist that you could spend your life looking at all of them. Photo by Gary A. Becker.

A Galaxy Filled with Stars

Astronomers classify stars in many ways, the most basic being by color. Stars are classed into spectral types along a continuum from blue to white to yellow to orange to red, as follows: O and B (blue), A and F (white), G (yellow), K (orange), and M (red). Color and temperature are directly related: the hotter the star, the bluer it is. Red stars are relatively cool. The Sun, as we've seen, lies in the middle as a yellowish G-type star.

You can see examples of the range of star colors with the naked eye on clear nights. Rigel (Beta Orionis), the 0-magnitude star marking Orion's left foot, is a B star. Nearby Sirius (Alpha Canis Majoris), the brightest star in the sky, is an A star. Zero-magnitude Procyon (Alpha Canis Minoris) is an F star.

THE BRIGHTEST STAR IN THE SKY, Sirius shines at magnitude -1.5, brighter than most of the planets. It is visible on winter evenings in the constellation Canis Major. Photo by Jim Baumgardt.

Zero-magnitude Capella (Alpha Aurigae) is a G star like our Sun. First-magnitude Aldebaran (Alpha Tauri) is a K star. First-magnitude Betelgeuse (Alpha Orionis) is an M star. A longstanding mnemonic helps astronomers remember the sequence of spectral letters: "O Be A Fine Girl (or Guy), Kiss Me!"

Many of the stars visible to the naked eye are nearby stars in our Galaxy, several hundred light-years away. All stars visible in backyard telescopes are members of our Galaxy. Astronomers further separate stars by their luminosity classes, which broadly define stars as supergiants, giants, or dwarfs. Most stars are dwarfs, or so-called main sequence stars. The Sun lies in the midst of the main sequence, with a diameter of 870,000 miles, an average luminosity, and a surface temperature of 5,800 K.

STARS CROWD TOGETHER toward the center of our Galaxy, in the direction of the constellation Sagittarius. All stars visible to the naked eye are members of our Milky Way Galaxy. Photo by Chris Schur.

At the extreme limits of a star's existence, some supergiants span hundreds of millions of miles, shine with the light of several thousand suns, and have surface temperatures of 30,000 K. White dwarfs are extremely dense, burned out stars that have collapsed on themselves and are no larger than Earth, emitting a tiny fraction of the Sun's light. These are different than red dwarfs, the normal dwarfs that lie on the main sequence.

Although the Sun lies in the middle of the scheme of things, slow-burning red dwarf stars vastly outnumber "middle of the road" stars, giants, and supergiants. Statistically, then, the Sun is a better than average star.

THE CLOSEST STAR beyond our Sun is Alpha Centauri (center), which lies 4.2 light-years away. The light we now see from the star has traveled through space for over 1,500 days. Photo by Ronald E. Royer.

Double and Multiple Stars

Although the Sun is a single star, a casual scan of the sky with a telescope will show that many stars seem to exist in pairs or multiple systems. Many of these objects are simply chance alignments of stars that actually lie at vastly different distances. These are called optical double stars. However, many stars do live in twin, triple, or quadruple systems that are physically bound by gravity and travel through space in orbit around a common center of mass. Astronomers call these objects binary stars and multiple stars. Curiously, bina-

SUPERNOVA 1987A (below center) burst forth in the Large Magellanic Cloud, a satellite galaxy of the Milky Way's, briefly outshining its host galaxy before fading away. Photo by Jim Barclay.

ry and multiple systems outnumber single stars in our Galaxy by a substantial fraction, making up about 60 percent of all observed stars.

Binary stars allow astronomers to study properties about stars that otherwise would be difficult to gauge. Having determined the spectral types of the stars, astronomers can watch the partners orbit each other and learn about the masses, distances, ages, and other information that adds to the knowledge bank about how stars form and evolve. Such systems offer a convenient laboratory for orbital mechanics that would not exist from observations of single stars or the Sun alone. Some binary systems offer the best evidence for the existence of exotic phenomena like black holes. It's impossible to see a black hole, but astronomers can see the effects such an object has on the surviving star.

Splitting Binaries with Telescopes

Backyard observers enjoy viewing double and multiple stars because of the pretty colors they sometimes display and the challenge of "splitting" the components. Hundreds of doubles are visible in even a 2-inch telescope, so they provide a nearly inexhaustible source of observing targets. And because they are bright points of light, many double and multiple stars can be observed from light-polluted areas with little problem.

Perhaps the most popular double star is Mizar (Zeta Ursae Majoris), the middle star in the handle of the Big Dipper. The magnitude 2.4 and 4.0 white components are separated by a generous 14.4". Albireo (Beta Cygni), the star marking the nose of the summertime swan, consists of a vivid yellow 3rd-magnitude primary star and a blue 5th-magnitude companion, separated by 34.3", an actual distance of 400 billion miles. Almach (Gamma Andromedae) is a superbly pretty binary with a 2nd-magnitude orange primary and 5th-magnitude blue secondary separated by 10". Eta Cassiopeiae is a fine double for small telescopes, consisting of magnitude 3.5 and 7.2 stars separated by 14". The star colors are yellow and pale red, respectively.

A DISTANT SUPERNOVA in the galaxy M58 in Virgo, visible just below the galaxy, was captured on film by many amateur astronomers in April 1988. Photo by Kim Zussman.

Variable Stars

The Sun seems like a reliable companion in our lives, constant in every way. Yet it varies ever so slightly in energy output. Although this variance is slight, other stars vary wildly in their light output for a variety of reasons. Astronomers classify these into a bewildering array of categories collectively called variable stars. Intrinsic variable stars vary in light output due to differences in the star itself; other types of variables fluctuate in brightness for external reasons. The most straightforward class of intrinsic variable is the pulsating variable, which varies due to physical expansion and contraction of the star. Like a swinging pendulum, pulsating variables bloat in size and then rebound by gravitational attraction. T Tauri stars, named after the prototype in the constellation Taurus, vary irregularly as their young atmospheres produce giant

DOUBLE STARS like Mizar in the constellation Ursa Major are visible by the hundreds in small telescopes. In fact, the majority of stars in our Galaxy are double or multiple systems. Photo by Tony Gondola.

NICKNAMED "THE WONDERFUL," the star Mira in Cetus varies over a large range of brightness with time. Many such variable stars are visible in binoculars and telescopes. Photo by Ronald E. Royer.

THE SUMMER MILKY WAY shows the great variety of stars that confronts the naked eye viewer on a moonless night. Photo by Mark Coco.

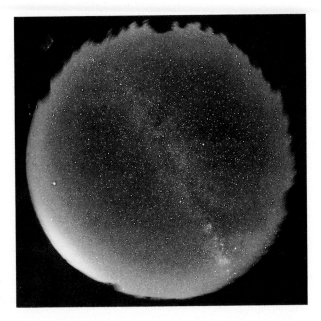

flares. Cepheid variable stars, named for the prototype Delta Cephei, show a definite relation between period and luminosity. Thus, they can be used as "standard candles" by astronomers who want to determine distances to objects that contain them.

Variables may have long or short periods of light fluctuation. RR Lyrae stars are short-period variables that fluctuate in brightness over intervals of less than one day. Long-period variables are aged, red stars like their prototype Mira (Omicron Ceti), that swing up and down in energy with periods of as long as several years. Eruptive variables and flare stars produce erratic surges in brightness. Eclipsing binaries drop and rise in brightness because one member of a double star system temporarily blocks the other. They are not intrinsic variables. Novae, a Latin word for new star, are stars in late stages of evolution that erupt dramatically in brightness. Supernovae, the most catalclysmic events in the universe, are high mass stars that run out of fuel to burn and explode, scattering debris throughout the space that surrounds them. Such supernovae can, for periods of a few days or weeks, outshine an entire galaxy.

10

The Young and the Aged:
A Look at Star Clusters

If you venture outside on a night when the Milky Way is visible, you'll see that stars in the glowing band of light are not uniformly distributed. Instead, you'll see areas of greater and lesser concentration — places where stars seem to clump together in groups separated by star-poor regions. Some of this clumpiness is due to a perspective effect, but in part the clumpiness is real. Stars and interstellar matter are not distributed evenly throughout the Galaxy. This is due to a variety of factors including the rotation of the Galaxy and the strong tug of gravitational attraction, but it underscores an important point about stars. They are not formed in isolation, but created together in star clusters.

Stars form from recycled hydrogen gas and dust that coalesces into a cloud, an emission nebula, that in effect is an interstellar nursery. Our Sun, for

THE PLEIADES (M45) cluster in Taurus is visible to the naked eye as a tiny, dipper-shaped clump of faint stars. Large scopes reveal faint blue nebulosity around several of the Pleiads. Photo by Bill and Sally Fletcher.

OBSERVERS GET A BONUS when viewing the sprawling open cluster M35 in Gemini. The more distant open cluster NGC 2158 lies in the background, and can be seen in the same field of view (below left). Photo by Preston Scott Justis.

example, was likely not always a lone star, but formed in a group within a nebulous cloud. If this is so, why don't all stars exist in clusters? Because the slow rotation of the Galaxy, acting over millions of years, slowly breaks apart star clusters and disperses their members throughout the spiral arms of the Galaxy. This is the scenario for the first of two major types of star clusters. The second type is a very different animal.

Open Star Clusters

Loose, irregularly shaped star clusters visible in the plane of the Milky Way are called open star clusters or sometimes galactic star clusters. These objects consist of young stars recently formed from a nebulous cloud of gas

THE WILD DUCK CLUSTER in Scutum lies in a rich Milky Way star field that is a pleasure to scan with binoculars. Photo by Bill Iburg.

and dust — in many cases, in fact, nebulosity not yet formed into stars is visible in association with the stars in the cluster. Open clusters exist in the disk portion of the Galaxy, and provide the means for the birth and early years of a huge percentage of stars. They generally contain a few hundred to a few thousand members, but can contain only a few dozen. They generally measure a few light-years across at most. All known open clusters are relatively close to the Sun — 10,000 light-years or less — because at greater distances open clusters merge into the rich background of individual Milky Way stars. Such clusters are important to astronomers because they formed at a given time in the history of the Galaxy and therefore allow a sampling of the chemical and astrophysical conditions present during the cluster's birth. The only significant variable in a star cluster is the range of masses for the various stars, an important point for astronomers wishing to gain insight on how stars burn their fuel and otherwise behave throughout their lives.

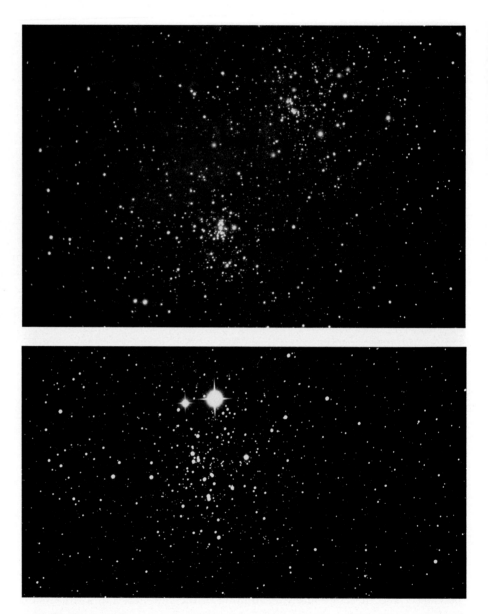

THE DOUBLE CLUSTER IN PERSEUS can be seen by the eye alone on any dark night. Photo by Bill Iburg.

THE OWL CLUSTER (NGC 457) in Cassiopeia somehow eluded inclusion in Messier's catalog despite its great brilliance in small telescopes. Photo by Preston Scott Justis.

Hunting Open Clusters with Binoculars

Because open star clusters lie relatively close to us in galactic terms, many consist of bright stars. A few dozen such clusters are beautiful objects not only in telescopes but binoculars, and several stunning examples can be seen with the eye alone under a dark sky. One of the most impressive such clusters is the Pleiades (M45) cluster in Taurus, a small, dipper-shaped group of bright stars also called the Seven Sisters. On dark winter nights the Pleiades is immediately visible to the naked eye as a tiny cloud of misty light. Look carefully and you'll see five or six tiny stars within the patch of light. Few people see seven stars with the eye alone — a test of keen eyesight. In binoculars, however, the Pleiades comes alive with dozens of gleaming blue and blue-white stars, many arranged in tight little geometric patterns. In a telescope, it's possible to

OMEGA CENTAURI is the greatest globular star cluster in the sky. Photo by Jack Newton.

A FAVORITE FOR NORTHERNERS, M13 in Hercules shows strongly defined lanes of stars that emanate from the cluster's core. Photo by Bill Iburg.

glimpse a small wisp of reflection nebulosity, dust associated with and illuminated by the young stars. The cluster is about 2° across — four times the size of the Moon — and lies at a distance of 400 light-years. Altogether the group shines with the light of a single 1st-magnitude star.

Other impressive open clusters for binoculars include the Hyades — located just south of the Pleiades — the great Coma Berenices star cluster (Melotte 111), The Beehive Cluster (M44) in Cancer, Southern Hemisphere wonders IC 2602 and NGC 3532, and a slew of clusters in the summer Milky Way that includes M6, M7, M39, and NGC 6231. Many fainter clusters are visible in telescopes, but you may find that you enjoy viewing the bright star groups the most in binoculars, whose wide fields enable you to include the entire cluster in a single field of view.

THE REGION AROUND ANTARES contains the bright globular M4 and a host of faint emission and reflection nebulae. Photo by Chris Cook.

Globular Star Clusters

The other type of star cluster is called a globular star cluster. Globular clusters are both astrophysically and observationally completely different from open clusters. They are giant spheres of stars many times larger than the biggest open clusters, lying outside the Galaxy's disk in the barren region called the galactic halo. They are old, predominantly yellow suns that formed long ago as the Galaxy settled into its present shape. Globular clusters are among the oldest unchanged entities in the universe, providing astronomers with a time capsule with which to view the Galaxy's early history. About 150 are known in the Milky Way, most lying from 10,000 to 200,000 light-years away, spanning 20 to 300 light-years across, and containing between 100,000 and 2 million stars.

Globular star clusters slowly orbit the center of the Galaxy in long, elliptical paths. Most of the time they are quiescent, far away from the action that permeates the Galaxy's disk. But every few tens of millions of years a globular cluster may pass down through the plane of the Galaxy and be swept clean of dust and gas that lies within it. Thus, globular clusters are essentially sterile, empty spheres containing only aged stars.

Resolving the Globes: Summertime Spectacle

Globular clusters are a joy to observe, and although at first glance many appear to be similar in photographs, they vary considerably when viewed with telescopes. Because they are distributed around the galactic center, most globular clusters appear to be concentrated in the summer sky (Sagittarius,

DISTANT GLOBULAR NGC 6440 lies adjacent to the challenging planetary nebula NGC 6445 (right). Photo by Jack Marling.

THE GLOBULAR CLUSTER M15 is a favorite object in the autumn sky. Photo by Bill Iburg.

the group of stars that marks the position of the galactic center, is a summer constellation). Because of this, few summertime observing events are more enjoyable than locating and viewing globular clusters with your telescope.

Several of the brightest globular clusters are visible — barely — to the naked eye. These include Omega Centauri (NGC 5139), the Hercules Cluster (M13), and clusters M4 in Scorpius and M22 in Sagittarius. But to see most globulars, even as fuzzy, bloated "stars," you need at least a good pair of binoculars. To resolve most globular clusters into their constituent stars, you'll need at least a 4-inch or 6-inch telescope.

The brightest globular cluster in the sky is Omega Centauri, visible as a 4th-magnitude "star" in the southern constellation Centaurus. Larger than the disk of the Moon, this hazy ball contains more than a million stars and is bright because it lies relatively close to the Sun. In a 6-inch telescope Omega Centauri appears like a huge hazy glow peppered with hundreds of tiny, white and yellow stars across an area spanning 30'. It is an amazing sight not easily forgotten.

The favorite globular cluster for Northern Hemisphere observers is the Hercules Cluster. Lying within the "keystone" asterism of Hercules, this 6th-magnitude object measures about half the diameter of Omega, and contains a more condensed, ball-like center. A 6-inch scope shows bright lanes of stars winding away from the center of this cluster, producing a vivid, three-dimensional effect. Other impressive globulars include M15 in Pegasus, M3 in Canes Venatici, 47 Tucanae in the deep southern sky, M5 in Serpens, and M55 in Sagittarius.

11

Cosmic Clouds:
Bright and Dark Nebulae

On a dark summer night one of the most awe-inspiring sights in the natural world arches overhead, sparkling and glistening like diamond dust on black velvet. Simply step outside on such a night and you'll see the Milky Way, the smooth, fuzzy glow of the disk of our Galaxy seen from within. It stretches from Scorpius and Sagittarius in the south, to Cygnus high overhead, to Cassiopeia in the north. Most of this glow comes from the unresolved light of countless thousands of stars. But scattered throughout the stars are thousands of nebulae — clouds of gas and dust — that litter the disk component of the Milky Way and play critical roles in the ways stars are born and die.

THE CORE OF THE ORION NEBULA (M42) consists of a thick blanket of emission nebulosity surrounding bright, young blue-white stars. Photo by Jim Baumgardt.

119

THE OMEGA NEBULA (M17) in Sagittarius is one of the highest surface-brightness nebulae in the sky, and is consequently a grand object for small scopes. Photo by Michael Stecker.

The primary constituent of these clouds is hydrogen, the most abundant element in the universe. Quantitatively hydrogen plays a crucial role in everything that exists in the Galaxy and beyond, including being a primary component of water, the lifegiving fluid so precious to Earth. Hydrogen and other elements compose these interstellar clouds, and the exact way they are composed, the way forces act on them, and where they exist in the Galaxy combine to characterize nebulae in different ways.

Emission Nebulae

Astronomers call the brightest and most easily observed interstellar clouds emission nebulae. These regions of gas glow like a soft light bulb by fluorescence because their atoms are excited into ionization by nearby hot stars. Many emission nebulae are interstellar birthplaces, regions where ancient gas

THE LAGOON NEBULA (M8) in Sagittarius offers the chance to see extensive nebulosity, a bright star cluster, and dark nebulae including a central "lagoon." Photo by Jack Newton.

121

HOT STARS lie embedded in the central portion of IC 1805, a giant emission nebula in Cassiopeia. Photo by Michael Stecker.

from dead stars has collapsed gravitationally to a point where new stars form from the recycled elements. When hot, young stars within emission nebulae throw off immense amounts of energy, the surrounding gas cloud lights up and is visible in telescopes hundreds of light-years away.

Other types of emission nebulae are included together with star-forming regions as a class because they glow in the same way, from radiation supplied by hot stars. Yet supernova remnants and planetary nebulae are very different animals from star-forming regions. Supernova remnants are the quickly expanding gaseous remains of massive stars that ended their lives in violent explosions after depleting their nuclear fuel. Planetary nebulae are the slowly expanding gaseous remnants of less massive stars that ended their existence by "belching" off layers of material after consuming their last stockpiles of nuclear fuel. Superb examples of all three types of emission nebulae are visible in backyard telescopes.

From Orion to Invisibility

The brightest emission nebulae are visible to the naked eye. Such is the case with the Orion Nebula (M42), the most famous of all emission regions, visible as a soft glow in the sword of Orion. This object is one of the largest emission nebulae known in our Galaxy, and is located only about 1,500 light-years away. Binoculars show that M42 is a nebular cloud peppered with a few bright stars lying inside it. A 4-inch telescope at moderate power shows distinct features in M42, such as the high surface brightness central core, the multiple star Theta^1 Orionis, double star Theta^2 Orionis, a dark protuberance called the

THE TRIFID NEBULA (M20) is composed of emission nebulosity, which shows up red, and reflection nebulosity, visible as strongly blue. Photo by Tony Hallas and Daphne Mount.

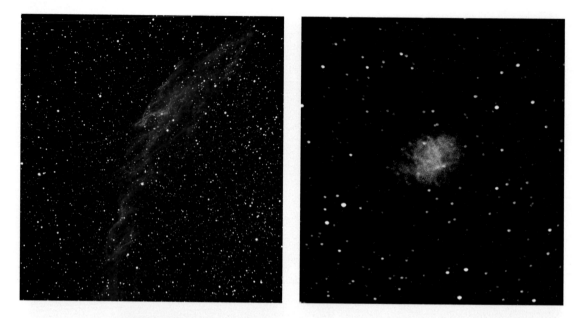

ONCE THOUGHT CHALLENGING for small scopes, the arc of the Veil Nebula, NGC 6992-5, is routine for an 8-inch scope under dark skies. Photo by Jack Newton.

THE CRAB NEBULA in Taurus is a bright supernova remnant, the exploded remains of a dead star. Photo by Jack Marling.

Fish's Mouth in the bright core, and a faint halo of nebulosity fanning away from the center. In total extent the Orion Nebula covers an area of 85' by 60', larger than the Full Moon, but only about half that area is visible through telescopes because of the faintness of the outlying parts of the nebula.

Other large, bright, and relatively close emission nebulae are visible to the naked eye under dark skies. These include the Eta Carinae Nebula (NGC 3372) — a magnificent object deep in the southern sky — the North America Nebula (NGC 7000) in Cygnus, IC 1396 in Cepheus, the Lagoon (M8) and Trifid (M20) nebulae in Sagittarius, the Eagle Nebula (M16) in Serpens, the Omega Nebula (M17) in Sagittarius, and the Tarantula Nebula (NGC 2070) in Dorado. Such emission nebulae aptly demonstrate the difference in sensitivity between photography and the human eye: although most look vivid red in long exposure photos, you'll see them in a telescope as gray-green. Unfortunately, eyes are not as sensitive to the same wavelengths of light as is film.

A handful of supernova remnants are visible in backyard telescopes. The reason for the relative rarity of these objects is simple. Over the span of several thousand years, for example, relatively few stars become supernovae in any single galaxy like the Milky Way. When a star does "go supernova," it produces a bright ring of gas that glows brightly and then slowly dims as it expands into the interstellar medium. Astronomers have calculated that the typical supernova remnant has a lifetime of only 20,000 to 50,000 years. So to see a supernova remnant as a reasonably bright object it must be relatively close in the Galaxy and less than 50,000 years old.

For small telescopes, the best example of a supernova remnant is the Crab Nebula (M1) in Taurus. This little oval nebulosity glows at about 9th magnitude and so is easy to see in a 4-inch telescope. Curiously, Chinese and Native American records describe the presence of a bright daytime star in the constellation Taurus that remained visible for several weeks in the year 1054. Indeed, this star was a supernova that produced the Crab Nebula and we are today, better than 900 years later, able to see the remains of this stellar explosion. Closeup photos of the Crab Nebula show thin tendrils of gas speeding out ahead of the cloud as it expands, and early large telescope observations of these tendrils produced the common name Crab Nebula. These features are difficult to see with anything but the largest backyard telescopes, however. Other notable sueprnova remnants for amateur scopes include the Veil Nebula (NGC 6960, NGC 6992-5) in Cygnus and the low-surface-brightness object NGC 6888 in Cygnus.

Dumbbells and Donuts

Because most are small and have relatively high surface brightnesses, planetary nebulae are much easier to observe than supernova remnants. The name planetary nebula derives from early observers who likened the telescopic appearances of these objects to the blue-green disks of the planets Uranus and Neptune. Although most planetaries appear like disks, they are actually spheres of gas and their uneven growth causes in some examples far more peculiar shapes than mere disks. The prototypical planetary is the Ring Nebula (M57)

THE BEST KNOWN DARK NEBULA is the Horsehead Nebula in Orion, visible with difficulty in large backyard telescopes. Photo by Barry Sobel.

in Lyra, an object measuring slightly over 1' across and possessing a dark center, such that it appears to be a ghostly smoke ring, or glowing donut.

The Dumbbell Nebula (M27) in Vulpecula is larger and brighter than the Ring. Given the inglorious name because its shape actually resembles a dumbbell, M27 is visible in binoculars as a fuzzy glow. Telescopes are required to show the shape distinctly, however, and a good 6-inch scope shows not only the dumbbell (or butterfly) shape but also fainter "ears" of nebulosity that nearly close the object into an illuminated oval. A good 8-inch telescope shows several stars splattered across the face of this nebula, the one in the center of the object — a dim, 13th-magnitude orb — being the progenitor star. The Dumbbell measures 8' by 6' in extent, large for a planetary nebula.

The Dumbbell is not the largest, brightest, or closest planetary nebula, however. That honor belongs to the Helical Nebula (NGC 7293) in Aquarius. Measuring 15' by 12', half the size of the Full Moon, the Helix is large enough to see easily in binoculars, but only from a dark sky. The nebula's total magnitude is 6.5, but it suffers from low surface brightness — its light is so spread out that little pieces of the nebula appear dim. Hence, a wide-field telescope at low power is the best instrument for viewing the elusive object. The Helical Nebula's southern declination places it low in the sky for many Northern Hemisphere viewers. Nevertheless, it is visible on dark, moonless nights.

Reflection Nebulae

Emission nebulae are not the only type of interstellar gas cloud, however. Reflection nebulae are lower energy clouds that have not been excited into luminescence. Rather, they are regions of dust and gas visible only because they reflect light from nearby stars. That a nebula is visible over hundreds of light-years simply by bouncing light from a nearby star is incredible, and as you might expect reflection nebulae are considerably fainter than many emission nebulae. The visibility of any nebula — and especially reflection objects — is controlled by the local geometry of the object and its position relative to neighboring stars.

Reflection nebulae are challenging observing targets; only a handful are bright enough to be easily viewed with small telescopes. One of the best is M78, a small gas cloud surrounding a pair of bright stars in central Orion, not far north of Orion's belt. Reflection nebulae have low surface brightnesses and M78 is no exception, so use a relatively wide-field eyepiece to find and observe M78. You'll see it as a faint glow surrounding the stars.

Substantially more challenging than M78 is the Merope Nebula (NGC 1435), an extremely faint patch of reflection nebulosity surrounding one of the brightest stars in the Pleiades (M45) star cluster in Taurus. The Merope Nebula is faint enough to require that you use a moderate size (8-inch or larger) telescope, averted vision, and place the star just outside of a high-power field to eliminate glare from the star. On nights of superb sky steadiness you may catch a glimpse of this deep-sky observing challenge.

Dark Nebulae

A third type of interstellar cloud is the hardest type to recognize. Dark nebulae are clouds of extremely fine dust particles not visible at all directly. To detect a dark nebula astronomers must see a lack of something else they know lies beyond it. For example, dark nebulae are most easily detected in the rich starfields toward the center of the Galaxy. In such fields patterns emerge where the background star density drops to nearly zero, approximating an opaque "hole" in the sky. Holes in the sky don't exist, despite the fact that early observational astronomers thought they might. The thousands of background stars do exist, so something must be blocking light from the background stars. The culprit? You guessed it — dark nebulae.

To say that observing dark nebulae is a challenge would be putting it mildly. To view a dark nebula you must observe something that is not there, or more properly something that is there but visible indirectly. Some dark nebulae are easier to see than others. The best type to start with are those that lie in front of extremely rich star fields — the greater the density of "missing" stars, the easier the dark nebula is to see. One such example is Barnard 86, one of an

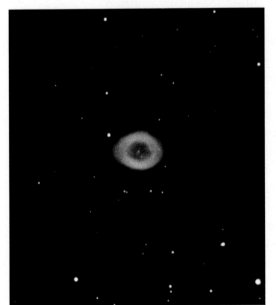

extensive catalog of dark nebulae compiled by the American astronomer E.E. Barnard. B86 is both easy to find and easy to see. It lies in star-rich Sagittarius next to the bright open cluster NGC 6520. A low-power telescope trained on this field on a dark night will most certainly show both the star cluster and the adjacent dark nebula, an extremely opaque "hole" with only a few foreground stars visible.

Other dark nebulae are far more challenging. The dark nebula catalogued as B72 and known as the Snake Nebula (or S Nebula) lies

THE RING NEBULA (M57) in Lyra is a cosmic puffball of subtly glowing gas. It is easy to find and is visible in any telescope as a small disk of light. Photo by Tony Hallas and Daphne Mount.

THE DUMBBELL NEBULA (M27) in Vulpecula is one of the most easily observed plane-taries in the sky. Photo by Jack Newton.

1.5° northeast of the bright star Theta Ophiuchi. Its winding, S-shaped form can be discerned on the darkest of nights. In the winter sky you will find the most famous dark nebula, the Horsehead Nebula (B33) in Orion. Lying just south of the belt star Zeta Orionis, the nebula is a shadowy form visible against the faint outline of emission nebula IC 434. The problem is not seeing the Horsehead, but the extremely faint emission nebula that backlights it. Use averted vision, moderately high power, and at least an 8-inch telescope to spot this legendary object.

The varied forms of nebulae offer some of the greatest sights of the night sky. From naked-eye glows to objects that need help to stand out against the blackness of space, the cosmic clouds offer hundreds of beautiful sights for the small telescope user. They are the most challenging telescopic sights of any component of our Galaxy. Only one realm of deep-sky object demands and offers more for backyard astronomers — the realm of the galaxies beyond our own.

12

A Universe of Galaxies

Thus far everything we've surveyed has been a component of the Milky Way Galaxy. Yet the universe is composed of billions of separate galaxies, each an independent, "island" star system. In terms of astronomical history, galaxies are new concepts. Throughout the hundreds of years of telescopic observation since 1610, astronomers observed many fuzzy patches in the sky, meticulously catalogued them, and labeled them "spiral nebulae." Yet they are completely different from nebulae, which we've seen are gas clouds scattered throughout galaxies like our Milky Way. Although galaxies appear similar to nebulae in earthbound telescopes, they are enormously larger and more distant than anything we've surveyed thus far.

Galaxies remained a mystery until 1923, when the American astronomer Edwin Hubble demonstrated the true nature of these cosmic beasts. Hubble identified a group of variable stars called Cepheids in one of the brightest "spiral nebulae," the Andromeda Galaxy, M31. The behavior of these Cepheids was precisely known because of the hundreds previously known in

SPIRAL GALAXY M81 in Ursa Major is an archetypal galaxy: a condensed nucleus, bright halo, and graceful spiral arms peppered with star-forming regions. Photo by Tony Hallas and Daphne Mount.

our Galaxy, yet Hubble found that these Cepheids were dramatically fainter than those that lay nearby. Hubble used the 100-inch Hooker telescope at the Mount Wilson Observatory near Los Angeles. He discovered that spiral nebulae were actually huge systems of stars incredibly far away, and the term galaxy was born.

What Makes a Galaxy?

Galaxies come in a multitude of shapes, masses, and luminosities. The Milky Way, a rather ordinary spiral galaxy, measures about 100,000 light-years from end to end. Yet some galaxies — the giant ellipticals — span 10 million light-years, while others — dwarf ellipticals — measure a relatively minute 10,000 light-years in diameter. A galaxy like ours contains about 200 billion stars. Its stellar population is dwarfed, however, by the trillions of stars in the giant ellipticals.

Generally, astronomers place galaxies into four broad classes. Spiral galaxies, of which our Milky Way and the Andromeda Galaxy are but two examples, consist of a spiral disk surrounded by a halo of gas and dust. Within the class of spiral galaxies, two subclasses exist: normal spirals and barred spirals. Barred spirals contain a distinct bar-shaped central region from which spiral arms emanate, but otherwise are similar to normal spirals.

Astronomers classify galaxies, including normal spirals (S) and barred spirals (SB) using a scheme devised by Hubble in the years following his discovery. The classifications Sa and SBa represent spirals and barred spirals with large centers and tightly wound spiral arms. Sb (and SBb) galaxies have smaller centers and spiral arms that are less tightly wound about the centers. Sc (and SBc) galaxies have tiny centers and large, patchy arms.

Elliptical galaxies are giant globes of stars and dust that lack the distinct disk shapes of spirals and seem to be organized only in the sense that they have concentrated nuclei. In the Hubble classification scheme, ellipticals are categorized from E0 to E7, with E0 representing a spherical galaxy and E7 an eccentric, oval object. Astronomers also recognize objects called lenticular galaxies that seem to consist of a spherical halo component like ellipticals yet contain a disk of stars as well.

THE ANDROMEDA GALAXY (M31) is the closest spiral galaxy to our Milky Way, and is the most distant object visible to human eyes. Photo by Tony Hallas and Daphne Mount.

THE CLASSIC EDGE-ON GALAXY is NGC 4565 in Coma Berenices, a silvery sliver of light visible on any dark night. Photo by Tony Hallas and Daphne Mount.

THE GALAXY M100 IN COMA BERENICES offered viewers a rare treat in 1979 by producing a bright supernova (visible as the brightest star near the galaxy's center). Photo by Bill Iburg.

Irregular galaxies are just that — irregular, and composed chiefly of discrete patches of stars and gas clouds. Their lack of form makes them as varied as any object in the universe, and they were the original "trash can" category devised by Hubble to facilitate classifying galaxies that didn't fit into the spiral or elliptical classifications. Another oddball category, peculiar galaxies, consists of distorted, interacting, or otherwise traumatically affected citizens of the cosmos.

The Local Group of Galaxies

Billions of galaxies are scattered throughout billions of light-years of the observable universe. Our little corner of the universe, however, constitutes what astronomers call the Local Group of Galaxies. Contained within a sphere approximately three million light-years in diameter, the Local Group is our home base in the galactic universe. It consists of the Milky Way and its companion galaxies, the Large and Small Magellanic Clouds, the Andromeda Galaxy and its retinue of small companions, and about two dozen other galaxies — mostly dwarf ellipticals. Other notable members include the Pinwheel Galaxy (M33) in Triangulum, Barnard's Galaxy (NGC 6822) in Sagittarius, Maffei 1 in Cassiopeia, the irregular galaxy IC 1613 in Sextans, and the dwarf galaxies in Fornax, Draco, Sculptor, Sextans, and Pegasus.

The most magnificent of the local galaxies is the Andromeda Galaxy, the sister spiral to our own. The Andromeda Galaxy is considerably larger than the Milky Way, and it is the only galaxy visible to the naked-eye from the Northern Hemisphere. Consult a star chart, go outside on a dark night, and look toward the right spot in Andromeda. You'll see the galaxy as a fuzzy, elongated patch of milky light. This galaxy is the most distant object visible to human eyes, lying approximately 2.2 million light-years away. It's a humbling thought to see the faint glow from this object and know that the photons passing into your eye began their journey through space as primitive mammals struggled to conquer the vast domain of Earth.

Galaxy Groups and Clusters

Just as our Milky Way belongs to a galaxy group, so do other galaxies. In fact on a larger scale galaxies exist in giant clusters and superclusters, and the wholesale structure of the universe at large is hardly homogeneous when viewed with backyard telescopes. As you begin to observe galaxies you'll note that they are not uniformly distributed across the sky. Part of this is due to the fact that dust in the plane of the Milky Way blocks light from galaxies beyond. However, the great windows into the deep universe visible in the spring and autumn evening skies demonstrate the clumping effect of galaxies near and far.

The greatest cluster of galaxies near us is the Virgo Cluster, visible as a collection of dozens of faint galaxies scattered over the constellations Virgo and Coma Berenices. Its most prominent members are M84, M86, and M87. Other outstanding and observable examples of galaxy clusters lie in the constellations Coma Berenices, Hercules, Fornax, Sculptor, and Perseus. Most galaxy groups and clusters visible in backyard telescopes lie tens of millions or even hundreds of millions of light-years away. The fact that we can see individual galaxies at such distances testifies to the immense energy output of these distant monsters of the deep.

Observational Tips for Galaxy Watchers

A small telescope is capable of showing thousands of galaxies. To see the most detail possible, however, you must observe from a dark sky site, and without the Moon in the sky. Take along a dimly lit red flashlight so you will not disturb your dark adaptation, your eyes' sensitivity to the galaxies' faint light. Make sure you have a star atlas sufficiently detailed to observe both bright and faint galaxies. If you wish, keep a journal of your observations or sketch what you see at the eyepiece.

Locate the galaxy you want to observe by star hopping, correlating the stars on your atlas to those you see in the sky. To find the galaxy, use a low-power eyepiece. When you center it in the field, study the galaxy and be sure to use averted vision — glancing to the side of the field. This uses your eyes' rods, its faint light detectors. As you look at the galaxy, make notes — mental or otherwise — on the galaxy's appearance. Is the center condensed and very bright? Can you see a faint halo of light surrounding the galaxy's center? Is spiral structure visible? Do you see other deep-sky objects in the field of view? After you've viewed the galaxy at low power, switch to a high magnification and see if more detail becomes visible. Before you know it, you'll be a seasoned galaxy watcher.

Many galaxies are not only observable

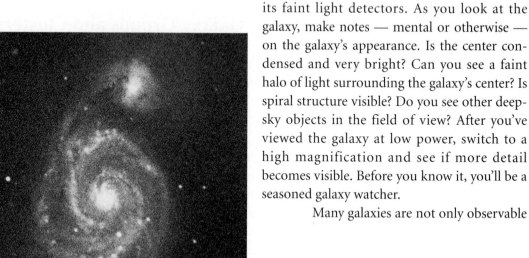

THE WHIRLPOOL GALAXY (M51) in Canes Venatici is a perennial favorite for galaxy watchers. The satellite galaxy, NGC 5195, is passing the Whirlpool itself. Photo by Martin C. Germano.

with small telescopes but show a reasonable amount of detail under dark skies. For Northern Hemisphere viewers, the Andromeda Galaxy and M33 are favorites. For Southern Hemisphere observers, the Magellanic Clouds offer unrivaled detail because these two Milky Way satellites lie relatively close to our own Galaxy.

Other galaxies worthy of early inspection include the big spiral NGC 253 in Sculptor, Centaurus A (NGC 5128) in Centaurus, the galaxy pair M81 and M82 in Ursa Major, the edge-on galaxy NGC 4565 in Coma Berenices, the Whirlpool Galaxy (M51) in Canes Venatici, M101 in Ursa Major, the Sombrero Galaxy (M104) in Virgo, NGC 7331 in Pegasus, M83 in Hydra, M63 and M94 in Canes Venatici, M100 in Coma Berenices, and NGC 5907 in Draco.

Centaurus A is a particularly outstanding object because it is so bright, large, and unbelievably weird. Although it has a far southern declination (and so is visible in the United States only from the southernmost states), it is well worth viewing if practicable. Carrying the number 5128 in the *New General Catalogue*, the galaxy picked up the alternate name Centaurus A because it is a strong emitter of radio energy. (The galaxy is also a powerful producer of X-ray and infrared emission.) Centaurus A's oddball structure, visible even in small scopes, demonstrates that something awfully peculiar is going on inside it. In the eyepiece Centaurus A appears as a glowing ball of light measuring some 18.2' by 14.5' in extent — nearly half the size of the Full Moon. The galaxy glows at magnitude 7.0, making it invisible to the unaided eye but visible as a faint smudge in binoculars.

In telescopes you'll see what appears to be a black band crossing the center of Centaurus A, causing it to appear bisected.

THE PECULIAR GALAXY CENTAURUS A (NGC 5128) consists of a huge spherical ball of stars, gas, and dust, encircled by a massive dust band. The galaxy is a bright object for binoculars. Photo by Steve Quirk.

THE ELLIPTICAL GALAXY NGC 404 lies within the same low-power field as bright star Beta Andromedae, making the galaxy difficult to spot. Photo by Jack Newton.

THE COMA BERENICES GALAXY CLUSTER holds dozens of distant galaxies visible in small scopes, many within a single field of view. The bright ellipticals are NGC 4889 and NGC 4874. Photo by Tony Hallas and Daphne Mount.

This in fact is a broad dust lane that encircles the galaxy and passes in front of the nucleus. High powers with large amateur telescopes — those 12 inches and larger in aperture — reveal faint glows within this dark dust band. These are huge chains of young, hot stars recently formed from the surrounding dust. Astronomers have long studied this galaxy as one of the most peculiar star systems known. It seems clear that because of its tremendous energy output, Centaurus A is home to a supermassive black hole that resides deep within the galaxy's nucleus.

Quasars

Galaxies have mysterious and extraordinarily distant cousins called quasars, an abbreviated name for quasi-stellar objects. Lying billions of light-years away, quasars are the energetic centers of extremely young galaxies. We see these powerful young objects because the farther we look back in space, the farther we are looking back in time. To see a quasar is to see a galaxy in its youthful chaos shortly after formation. To see an object at such a large distance, it must be incredibly bright. The energy output of a quasar exceeds that of anything else in the universe, and astronomers believe supermassive black holes lie deep inside these objects and power them like central engines.

A few dozen quasars are visible in backyard telescopes, but because of their great distance they appear as faint stars. The brightest quasar is 3C 273, located in the constellation Virgo. This object glows feebly at magnitude 12, and is slightly variable in light output. Quasars are to deep-sky objects what Pluto is to the solar system — a joy to spot not for the beautiful detail they show but for the euphoria of having looked across the universe.

LACKING STRUCTURE ENTIRELY, the unusual galaxy NGC 4038-9, nicknamed the Ringtail, is one of the great sights in Corvus. Photo by Tony Hallas and Daphne Mount.

THE BLACKEYE GALAXY (M64) in Coma Berenices shows a distinctive dark patch near its center. Photo by Bill Iburg.

139

13

Where Is Amateur Astronomy Going?

As a practice, amateur astronomy is in its infancy. Although beings have looked skyward for millions of years, the invention of the telescope for astronomical purposes occurred in 1609, less than 400 years ago. The instrument crucial to detailed astronomical observing remained firmly in the hands of scientists until quite recently. Amateur astronomy would not rise up as a movement in the United States until a rather peculiar period for societal functions — the Civil War. In 1862, as McClellan's army slowly tramped toward Richmond, a small group of interested persons formed the Chicago Astronomical Society. It is the oldest such group in the U.S., and it began a movement that has seen hundreds of astronomy societies come and go. Today some 250 astronomical societies exist in the U.S. and Canada, forming the backbone of a community of enthusiasts who actively observe, study, and think about the universe.

AN AURORA BLANKETS THE SKY above a backyard observatory in Pennsylvania. Will amateur astronomers always be able to observe the sky from convenient locations? Photo by Gary A. Becker.

PEERING INTO THE EYEPIECE, California amateur astronomer Ronald E. Royer prepares to observe a solar eclipse. Sketching a comet in 1946 started Royer's lifelong interest in astronomy. Photo courtesy Ronald E. Royer.

Amateur astronomy got such a slow start substantially because of the difficulty of obtaining telescopes. In fact, before World War II, amateur astronomers had to make their own optics for telescopes. As far back as the 1920s astronomers and enthusiasts like Russell Porter and Albert G. Ingalls promoted telescope making in places such as Scientific American. However, unless amateur astronomers first became amateur telescope makers, they were essentially excluded from easily observing the universe. This began to change by the middle of the century. Availability of surplus optics following World War II and a burgeoning telescope industry spurred by the advent of the "space age" resulted in more affordable and easier-to-use telescopes by the 1960s. Today, telescopes are better and more easily obtained than ever, and the telescope industry continues to produce a quality product at a reasonable price. In fact, most scientific instruments are fantastically more expensive than telescopes.

The Imaging Revolution

In the early 1990s the greatest excitement in amateur astronomy is coming from an imaging revolution. Electronic charge-coupled devices (CCDs) enable backyard astronomers to capture images of faint objects quickly and effectively and without the old-fashioned darkroom. Today the darkroom is replaced by a computer, and the practice of recording the sky on a permanent medium is undergoing a revolution.

THE QUEST FOR MARS has led NASA to launch a Mars Observer mission and anticipate a manned mission to the Red Planet. Such solar system exploration will invariably provide amateur astronomy with a boost of public interest. NASA photo.

Some years ago CCDs were frightfully expensive. The CCD chip is now a commonplace item in many pieces of consumer electronics — chiefly camcorders — and is therefore cheap enough to make CCD cameras for astronomy affordable to many amateur astronomers. A CCD is a modern replacement for photographic film: rather than storing photons in exposed grains of silver nitrate, a CCD stores photons digitally by turning on picture elements, or "pixels," struck by light. The CCD's efficiency of recording faint light is far higher than that of film, so exposures are much shorter, and the tricky aspects of astrophotography like guiding and alignment are less problematic. Typical exposures with a CCD camera are on the order of one to ten minutes rather than thirty minutes to two hours. Exposures of bright objects like planets are shorter yet. Slowly but surely more and more CCD imaging setups are showing up at star parties, complete with portable computers set up in the field. This growing interest is certain to continue over the coming years as CCDs themselves continue to become better and cheaper.

Astrophotography

Even now, however, good old-fashioned astrophotography is hardly dead. On the contrary, despite the interest in electronic imaging, there seems to be a resurgence of interest in conventional astrophotography. This is clearly evidenced by the increasing numbers of photo submissions — many by first-time photographers — to ASTRONOMY magazine.

POSED WITH HIS FIRST TELESCOPE, planetary observer Jim Phillips of Charleston, South Carolina, smiles with joy in a photo taken in September 1965. Today telescopes are far more powerful and much easier to obtain. Photo courtesy Jim Phillips.

Several factors are keeping conventional astrophotography a major part of amateur astronomy. First, CCD imaging is not a direct replacement, at least in terms of current technology. Some photographers like the results from film, whether it be slides or prints. Second, CCD imaging is still relatively expensive because it involves owning a computer loaded with memory for image processing. But the main reasons for the continuing popularity of astrophotography come from trends in the photo industry.

During the 1980s the major photo companies released a new series of films that were much better for astrophotography in two ways. These films provided far greater speed than earlier emulsions, cutting down significantly on the exposure times necessary to record faint objects. They also offered much finer resolution, making photos of everything from the Moon to galaxies a far more appealing product than films of the 1970s. As I write this book still more advanced films are getting ready to hit the market, and undoubtedly every year or two the "hot" films for astrophotography will change.

Because of the great speed and quality of the new generations of films astrophotography at a basic level is now very easy and inviting. By placing a camera loaded with fast color film on a sturdy tripod, a first-time photographer can shoot beautiful "nightscapes" — constellation, star trail, meteor, or aurora photos with horizon trees and landscape — the first night out. All such photos require are a 20- to 40-second exposure. The new generation of films makes through-the-telescope photography, with its pitfalls of alignment, focusing, and guiding, far easier than it was a generation ago.

The Deep Sky Crowd

For every backyard astronomer who records the sky on film, however, there are 100 who simply look at the objects. Given a little background knowledge, soaking in the light from a double star can be an inspiring event. "Hey, did you know the inky black gap of space between those two stars in the eyepiece is just about 400 billion miles?" Such are the things that intrigue beginners and bring the intellectual power of knowing the universe home to seasoned skywatchers. The group that pushes their knowledge of the sky and what it contains to the limits is the largest specialty "fraternity" in amateur astronomy — the deep-sky observers.

The past ten or twenty years have witnessed a revolution in deep-sky observing. With the mass-produced, easily available large aperture telescopes, thousands of clusters, nebulae, and galaxies are now visible in the average tele-

THE FUTURE OF ASTROPHOTOGRAPHY is secure as films and cameras are becoming increasingly better, believes San Antonio astrophotographer Robert Reeves. Photo by Alan Sifford.

WILL THE MILKY WAY BE VISIBLE through light pollution fifty years from now? It's a question that California astrophotographers Bob Provin and Brad Wallis feel strongly about: "We may have to move farther from the overgrown metropolis to find a sky suitable for astronomy." Photo by Alfred Lilge.

scope. Just as important has been the revolution in information for deep-sky observers. Twenty years ago popular journals claimed that objects like the Veil Nebula, a supernova remnant in Cygnus, was "beyond the reach of amateur instruments." The expertise, persistence, and publishing efforts of the current generation of deep-sky observers have demonstrated that few objects are beyond the reach of large amateur scopes.

Because of the lure of faint and exotic objects, deep-sky observing has come to be the dominant activity within amateur astronomy. Larger and better telescopes and more sophisticated and reliable literature will certainly continue to be produced, continuing the evolution of peering deep into the Galaxy and the universe. However, deep-sky observing faces one threat: light pollution. The growing regions of nighttime lighting aimed skyward threaten the ability of amateur astronomers to observe from dark skies, a necessity for good deep-sky viewing. Technology has helped somewhat in this area by making possible wavelength-specific nebula filters that, placed in your telescope, block the wavelengths of manmade lighting while allowing the light from a nebula or galaxy to pass into your eyepiece. These accessories work well but can hardly reverse the growing problem of bright skies at night.

Star Parties

Of all the events associated with amateur astronomy the most treasured are annual star parties, gatherings where people come together to share their common interest. Several decades ago only one large star party existed. Today dozens of these events — some on a national level — vie for pieces of the summertime when vacation time is available. The "big three" star parties in the United States are the Texas Star Party (TSP), held in the springtime near Fort Davis, Texas; Stellafane, the oldest star party, held in midsummer in Springfield, Vermont; and the Riverside Telescope Makers Conference (RTMC), held each Memorial Day weekend at Big Bear, California.

Each star party has a distinctly different character. The TSP hosts about 800 of the most active observers in the country, many equipped with 18-, 20-, or even 25-inch Dobsonian reflectors they have brought to the Prude Ranch several miles from McDonald Observatory. The emphasis here is on serious

TELESCOPE MAKING allows one to assemble a custom-made instrument for the same amount of money as a commercial scope. Susan French's reflector shows great pride in workmanship. Photo courtesy Susan French.

GREAT AMATEUR ASTRONOMERS like Frenchman Jean Dragesco often happen into an interest in the sky. "I discovered both microscopes and telescopes at an early age," he says, "and after tinkering with lenses found I loved viewing the sky." Photo courtesy Jean Dragesco.

observing, and a walk through the field on a clear evening reveals star atlases illuminated by dim red flashlights, computer screens weakly glowing by phosphor, and shouts and whispers about various NGC, IC, UGC, MCG, or Terzan objects. Many of the TSP regulars are experts who go after the most challenging deep-sky objects visible in transportable telescopes. The TSP rolls on with talks, tours of McDonald Observatory, and all-night observing for an entire week.

The outgrowth of a small group of telescope builders who formed a club in the 1920s, Stellafane is as much a tradition as it is an observing event. About 1,500 people typically converge on Breezy Hill in Springfield each July or August and the central focus here is not on the sky but on the telescopes. Annual telescope judging takes place and entrants receive awards for mechanical ingenuity, optical quality, and overall craftsmanship. The skill and dedication visible in the many scopes set up each year on the hill near Stellafane's historic little pink clubhouse demonstrate that the art of telescope making is very much alive.

The Riverside Telescope Makers Conference also emphasizes telescope construction, although when conditions permit the attendees do a fair amount of observing here too. In recent years the southern California crowd, which has swelled to 2,000 during the busiest years, has had the opportunity to see commercial telescopes set up by manufacturers to purchase discounted merchandise. The affair is as much a telescope show as a telescope maker's gathering.

Throughout the 1970s and 1980s these three events dominated the national amateur astronomy scene. Now there is an explosion of interest and activity in astronomy events, and many other star parties can boast regular attendance of several hundred. Attending one of these events is one of the best ways to see what's happening in the hobby and to get in touch with other amateur astronomers with interests similar to yours. Among the notable events are Astrofest (sponsored by the Chicago Astronomical Society and held in Kankakee, Illinois), the Apollo Rendezvous (Miami Valley Astronomical Society, Dayton, Ohio), the Okie-Tex Star Party (Oklahoma City Astronomy Club, Fort Davis, Texas), Astronomy Jamboree (Custer Institute Observatory, Southold, New York), the Winter Star Party (Southern Cross Astronomical Society, Florida Keys, Florida), Hidden Hollow (Richland Astronomical Society, Mansfield, Ohio), Mount Kobau Star Party (Osoyoos, British Columbia), and gatherings sponsored in different cities each year by the Astronomical Society of the Pacific and the Astronomical League.

STAR PARTIES are gatherings of amateur astronomers that focus on observing but allow participants in the hobby to socialize, give talks, and share their common interest. Photo courtesy Jacques Guertin.

149

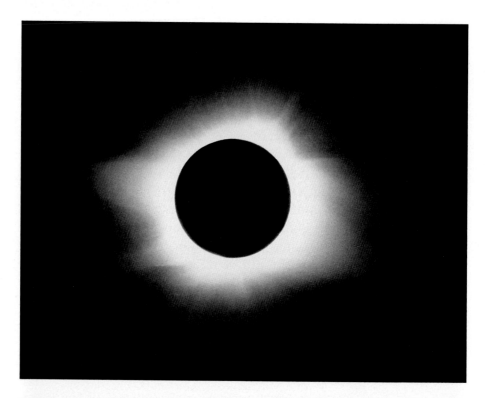

WHERE IS AMATEUR ASTRONOMY GOING? Will the field blossom over the coming decades or be eclipsed by other activities? Nearly all amateur astronomers show a healthy optimism for growing public interest and activity within the community of amateur astronomy. Photo by Andrew Steinbrecher.

If you can't easily make it to a star party, seek out an astronomy club. (See Appendix 4 for a listing of North American astronomy clubs.) By attending a few club meetings, talking with some members who have been involved with the hobby for a few years, and perhaps taking some peeks through the eyepiece at a local club stargaze, you'll get a good idea of what most interests you in the field. You can also get expert advice on what types of telescopes might be best for you. It's good to enter the hobby with as much information as you can get, and astronomy is like no other activity in that respect. Astronomy, however, can promise you the Moon and stars and never go back on its word. Happy observing!

Appendix 1: The Solar System

Object	Semi-major axis	Diameter	Mass	Rev. Period	Rot. Period	Atmosphere
Sun	—	1,392,000 km	1.99×10^{30} kg	—	25.4 days	hydrogen, helium
Mercury	0.387 AU	4,878 km 3.30×10^{23} kg	3.30×10^{23} kg	87.97 days	58.65 days	none
Venus	0.723 AU	12,104 km	4.87×10^{24} kg	224.70 days	243.0 days	carbon dioxide
Earth	1.000 AU	12,756 km	5.98×10^{24} kg	365.26 days	1.0 day	nitrogen, oxygen
(Moon: The Moon.)						
Mars	1.524 AU	6,787 km	6.42×10^{23} kg	686.98 days	1.026 days	carbon dioxide
(Moons: Phobos, Deimos.)						
Jupiter	5.203 AU	142,980 km	1.90×10^{27} kg	11.86 years	0.410 days	hydrogen, helium
(Moons: Metis, Adrastea, Amalthea, Thebe, Io, Europa, Ganymede, Callisto, Leda, Himalia, Lysithea, Elara, Ananke, Carme, Pasiphae, Sinope.)						
Saturn	9.539 AU	120,540 km	5.69×10^{26} kg	29.46 years	0.444 days	hydrogen, helium
(Moons: Pan, Atlas, Prometheus, Pandora, Janus, Epimetheus, Mimas, Enceladus, Tethys, Telesto, Calypso, Dione, Helene, Rhea, Titan, Hyperion, Iapetus, Phoebe.)						
Uranus	19.182 AU	52,120 km	8.6×10^{25} kg	84.01 years	0.718 days	hydrogen, helium
(Moons: Cordelia, Ophelia, Bianca, Cressida, Desdemona, Juliet, Portia, Rosalind, Belinda, Puck, Miranda, Ariel, Umbriel, Titania, Oberon.)						
Neptune	30.058 AU	49,530 km	1.03×10^{26} kg	164.79 years	0.671 days	hydrogen, helium
(Moons: Naiad, Thalassa, Despina, Galatea, Larissa, Proteus, Triton, Nereid.)						
Pluto	39.44 AU	2,300 km?	9×10^{21} kg ?	247.69 years	6.39 days ?	methane
(Moon: Charon.)						

Appendix 2: The Messier Catalog

Object	Popular name	Type	Constellation	Object	Popular name	Type	Constellation
M1 (NGC 1952)	Crab Nebula	SNR	Taurus	M58 (NGC 4579)	—	G	Virgo
M2 (NGC 7089)	—	GC	Aquarius	M59 (NGC 4621)	—	G	Virgo
M3 (NGC 5272)	—	GC	Canes Venatici	M60 (NGC 4649)	—	G	Virgo
M4 (NGC 6121)	—	GC	Scorpius	M61 (NGC 4303)	—	G	Virgo
M5 (NGC 5904)	—	GC	Serpens	M62 (NGC 6266)	—	GC	Ophiuchus
M6 (NGC 6405)	Butterfly Cluster	OC	Scorpius	M63 (NGC 5055)	—	G	Canes Venatici
M7 (NGC 6475)	—	OC	Scorpius	M64 (NGC 4826)	Blackeye Galaxy	G	Coma Berenices
M8 (NGC 6523)	Lagoon Nebula	EN+OC	Sagittarius				
M9 (NGC 6333)	—	GC	Ophiuchus	M65 (NGC 3623)	—	G	Leo
M10 (NGC 6254)	—	GC	Ophiuchus	M66 (NGC 3627)	—	G	Leo
M11 (NGC 6705)	Wild Duck Cluster	OC	Scutum	M67 (NGC 2682)	—	OC	Cancer
M12 (NGC 6218)	—	GC	Ophiuchus	M68 (NGC 4590)	—	GC	Hydra
M13 (NGC 6205)	Hercules Cluster	GC	Hercules	M69 (NGC 6637)	—	GC	Sagittarius
M14 (NGC 6402)	—	GC	Ophiuchus	M70 (NGC 6681)	—	GC	Sagittarius
M15 (NGC 7078)	—	GC	Pegasus	M71 (NGC 6838)	—	GC	Sagitta
M16 (NGC 6611)	Eagle Nebula	EN+OC	Serpens	M72 (NGC 6981)	—	GC	Aquarius
M17 (NGC 6618)	Omega Nebula	EN	Sagittarius	M73 (NGC 6994)	—	OC	Aquarius
M18 (NGC 6613)	—	OC	Sagittarius	M74 (NGC 628)	—	G	Pisces
M19 (NGC 6273)	—	GC	Ophiuchus	M75 (NGC 6864)	—	GC	Sagittarius
M20 (NGC 6514)	Trifid Nebula	EN	Sagittarius	M76 (NGC 650-1)	Little Dumbbell	PN	Perseus
M21 (NGC 6531)	—	OC	Sagittarius	M77 (NGC 1068)	—	G	Cetus
M22 (NGC 6522)	—	GC	Sagittarius	M78 (NGC 2068)	—	RN	Orion
M23 (NGC 6494)	—	OC	Sagittarius	M79 (NGC 1904)	—	GC	Lepus
M24	Sgr. Star Cloud	SC	Sagittarius	M80 (NGC 6093)	—	GC	Scorpius
M25 (IC 4725)	—	OC	Sagittarius	M81 (NGC 3031)	—	G	Ursa Major
M26 (NGC 6694)	—	OC	Scutum	M82 (NGC 3034)	—	G	Ursa Major
M27 (NGC 6853)	Dumbbell Nebula	PN	Vulpecula	M83 (NGC 5236)	—	G	Hydra
M28 (NGC 6626)	—	GC	Sagittarius	M84 (NGC 4374)	—	G	Virgo
M29 (NGC 6913)	—	OC	Cygnus	M85 (NGC 4382)	—	G	Virgo
M30 (NGC 7099)	—	GC	Capricornus	M86 (NGC 4406)	—	G	Virgo
M31 (NGC 224)	Andromeda Galaxy	G	Andromeda	M87 (NGC 4486)	—	G	Virgo
M32 (NGC 221)	—	G	Andromeda	M88 (NGC 4501)	—	G	Coma Berenices
M33 (NGC 598)	Pinwheel Galaxy	G	Triangulum				
M34 (NGC 1039)	—	OC	Perseus	M89 (NGC 4552)	—	G	Virgo
M35 (NGC 2168)	—	OC	Gemini	M90 (NGC 4569)	—	G	Virgo
M36 (NGC 1960)	—	OC	Auriga	M92 (NGC 6341)	—	GC	Hercules
M37 (NGC 2099)	—	OC	Auriga	M93 (NGC 2447)	—	OC	Puppis
M38 (NGC 1912)	—	OC	Auriga	M94 (NGC 4736)	—	G	Canes Venatici
M39 (NGC 7092)	—	OC	Cygnus				
M41 (NGC 2287)	—	OC	Canis Major	M95 (NGC 3351)	—	G	Leo
M42 (NGC 1976)	Orion Nebula	EN	Orion	M96 (NGC 3368)	—	G	Leo
M43 (NGC 1982)	—	EN	Orion	M97 (NGC 3587)	Owl Nebula	PN	Ursa Major
M44 (NGC 2632)	Beehive Cluster	OC	Cancer	M98 (NGC 4192)	—	G	Coma Berenices
M45 (Melotte 22)	Pleiades Cluster	OC	Taurus				
M46 (NGC 2437)	—	OC	Puppis	M99 (NGC 4254)	—	G	Coma Berenices
M47 (NGC 2422)	—	OC	Puppis				
M48 (NGC 2548)	—	OC	Hydra	M100 (NGC 4321)	—	G	Coma Berenices
M49 (NGC 4472)	—	G	Virgo				
M50 (NGC 2323)	—	OC	Monoceros	M101 (NGC 5457)	—	G	Ursa Major
M51 (NGC 5194)	Whirlpool Galaxy	G	Canes Venatici	M102 (NGC 5866)	—	G	Draco
				M103 (NGC 581)	—	OC	Cassiopeia
M52 (NGC 7654)	—	OC	Cassiopeia	M104 (NGC 4594)	Sombrero Galaxy	G	Virgo
M53 (NGC 5024)	—	GC	Coma Berenices	M105 (NGC 3379)	—	G	Leo
				M106 (NGC 4258)	—	G	Canes Venatici
M54 (NGC 6715)	—	GC	Sagittarius				
M55 (NGC 6809)	—	GC	Sagittarius	M107 (NGC 6171)	—	GC	Ophiuchus
M56 (NGC 6779)	—	GC	Lyra	M108 (NGC 3556)	—	G	Ursa Major
M57 (NGC 6720)	Ring Nebula	PN	Lyra	M109 (NGC 3992)	—	G	Ursa Major

The Messier Catalog, compiled by French comet hunter Charles Messier between 1758 and 1783 (and extended for years afterward by his contemporaries), is generally held as the most used list of bright deep-sky objects. The list was originally compiled to mark fixed nebulous objects that should not be confused with comets. Soon, however, the list became the starting point for great explorations of what astronomers would later discover were star clusters, nebulae, and galaxies. Object type code: EN = emission nebula; G = galaxy; GC = globular star cluster; OC = open star cluster; PN = planetary nebula; RN = reflection nebula; SNR = supernova remnant.

Appendix 3: Telescope Manufacturers

The following lists major telescope manufacturers who can provide literature on their product lines. Most produce not only telescopes, but extensive lines of accessories like eyepieces, binoculars, filters, mounts, and cameras.

Astro-Physics,
11250 Forest Hills Road,
Rockford, Illinois 61111.
(815) 282-1513.

Celestron International,
2835 Columbia Street,
Torrance, California 90503.
(310) 328-9560. FAX (310) 212-5835

Coulter Optical, Inc.,
P.O. Box K,
Idyllwild, California 92349
(714) 659-4621

Edmund Scientific Co.,
Edscorp Bldg.,
Barrington, New Jersey 08007
(609) 573-6250. FAX (609) 573-6295

Fujinon, Inc.,
10 High Point Drive,
Wayne, New Jersey 07470
(201) 633-5600

Lumicon,
2111 Research Drive #5,
Livermore, California 94550
(510) 447-9570 FAX (510) 447-9589

Meade Instruments Corporation,
1675 Toronto Way,
Costa Mesa, California 92626
(714) 556-2291 FAX (714) 556-4604

Orion Telescope Center,
2450 17th Avenue,
Santa Cruz, California 95061
(408) 464-0446

Parks Optical,
270 Easy Street,
Simi Valley, California 93065
(805) 522-6722

Questar Corporation,
P.O. Box 59,
New Hope, Pennsylvania 18938
(215) 862-5277

Swift Instruments, Inc.,
952 Dorchester Avenue,
Boston, Massachusetts 02125

Tele Vue Optics,
20 Dexter Plaza,
Pearl River, New York 10965
(914) 735-4044

Texas Nautical Repair Company,
3110 South Shepard,
Houston, Texas 77098
(713) 529-3551 FAX (713) 529-3108

Thousand Oaks Optical,
P.O. Box 248098,
Farmington, Michigan 48332
(313) 353-6825 FAX (313) 350-3111

Unitron, Inc.,
175 Express Street,
Plainview, New York 11803
(516) 822-4601

Appendix 4: North American Astronomy Clubs

ALABAMA

Auburn. Auburn Astro. Soc., 748 E. Magnolia Ave., 36830, Jim Chestnutt, (205) 821-1062.
Birmingham. Birmingham Astro. Soc., Box 36311, 35236, Doug Finch, (205) 956-2818.
Huntsville. Von Braun Astro. Soc., Box 1142, 35807.
Mobile. Mobile Astro. Soc., Box 169972, 36616-1972, Leland Cox, (205) 633-9079.
Rainbow City. Southside Astro. Club, 103 Eddie Circle, 35901, Julian Long, (205) 442-7169.
Tuscumbia. Muscle Shoals Astro. Soc., 302 Northeast Commons, 35674, Homer Russell, Jr., (205) 383-6717.

ALASKA

Anchorage. Anchorage Astro. Soc., 3539 Dunkirk Dr., 99502, John Morrow, (907) 269-5744.
Fairbanks. Astro. Units, Box 82210, 99708, Robert E. Fischer, (907) 456-6586.

AMERICAN SAMOA

Pago Pago. American Samoa Astro. Soc., Box 3076, 96799.

ARIZONA

Mesa. Leisure World Astron., 1792 Leisure World, 85206, Bud Ridley, (602) 981-3136.
Phoenix. Saguaro Astro. Club, Paul Lind, (602) 863-3077.
Prescott. Prescott Astro. Club, 1848 Emerald Dr., 86301.
Scottsdale. Phoenix Astro. Soc., 6945 East Gary Road, 85254, Raul V. Espinoza, (602) 996-3617.
Sierra Vista. Huachuca Astro. Club, 11 E. Berridge Dr., 85635, Theodore Nelson, (602) 458-5973. Founded 1981.
Sun City West. Astro. Club of Sun City West, 17802 N. 131st Ave., 85375, Jim Crisman, (602) 584-0896.
Tucson. Tucson Amateur Astro. Assoc., 10385 Observatory Dr., 85747, Derald Nye.

ARKANSAS

Arkadelphia. Ark-La-Tex Skywatchers, Box 493, 71923, Rex McDaniel.
Dover. Arkansas Amateur Astron., 20 Rolling Green Dr., 72837, Rex McDaniel, (501) 968-8772.
Fayetteville. Astro. Soc. of Northwestern Arkansas, 2503 Sweetbriar Dr., 72703.
Hackett. Arkansas-Oklahoma Astro. Soc., Coleman Memorial Observatory, Box 404, Route 1, 72937, (501) 474-4740.
Little Rock. Mid-South Astro. Soc., Box 5142, 72225, James Bruce McMath, (501) 376-3021.
Texarkana. Red River Astro. Club, 4808 Sunbury, 75502, Ken White, (501) 772-4200.

CALIFORNIA

Aptos. Cabrillo Astro. Stargazing Soc., Cabrillo College, 6500 Soquel Dr., 95003.
Arcata. Arcata Soc. for Amateur Astron., 3211 Alice Ave., 95521.
Bakersfield. Kern Astro. Soc., 4209 Tyndall Ave., 93313, David Hale.
Berkeley. Sidewalk Astron., Box 10003, 94709, Gerald Pardeilhan, (415) 527- 4026.
Big Bear Lake. Bear Valley Astro. Soc., Box 874, 92314.
Chatsworth. Celestial Observers, 9534 Gierson Ave., 91311-4737, Michael Selden, (818) 882-6172.

Costa Mesa. Orange County Astron., 2195 Raleigh Ave., 92627, John Sanford,   (714) 722-7900.
Cucamonga. Pomona Valley Amateur Astron., 8773 Balsa St., 91730, Bob Hibble, (714) 989-3680.
Eureka. Astron. of Humboldt, Box 351, 95502, Robert Zigler, (707) 768-3296.
Fresno. Central Valley Astron., 5790 East Tarpey Ave., 93727, Clarence Funk, (209) 291-7879.
Hemet. Hemet Astro. Soc., 561 West Devonshire Ave., 92343, Ralph L. Shook, (714) 658-3631.
Idyllwild. Idyll-Gazers Amateur Astro. Club, Box 1245, 92349, Joe Neu, (714) 659-3562.
Joshua Tree. Andromeda Astro. Soc., Star Route 1, Box 1112, 92252, Bill Crawford.
Lakewood. Excelsior Telescope Club, St. Pancratius Catholic Church, 3519 St. Pancratius Place, 90712, Ronald Royer, (213) 634-6111.
Lancaster. Antelope Valley Astro. Soc., Box 426, 93534, Steve Mathis, (805) 256-4261.
Los Angeles. Los Angeles Astro. Soc., 2800 East Observatory Road, 90027, Timothy Thompson, (213) 833-1733.
Mill Valley. Marin Stargazers, 27 Morning Sun, 94941.
Modesto. Stanislaus Amateur Astron., 1521 Clevenger Dr., 95356-0810, James Brockway, (209) 522-0082.
Moraga. Astro. Assoc. of Northern California, 731 Camino Ricardo, 94556, Don Stone, (415) 376-3007.
Mountain View. Peninsula Astro. Soc., Box 4542, 94040, Bill Sorrells, (415) 566-3116.
Murrieta. Temecula Valley Astron., 37795 Sea Pines Court, 92362, Kent Smith, (714) 698-0975.
Newhall. Local Group of Santa Clarita Valley, 25032 Walnut St., 91321, LaVerne Booth, (805) 259-3284.
Oakland. Chabot Telescope Makers Workshop, 4917 Mountain Blvd., 94619.
Oakland. Eastbay Astro. Soc., 4917 Mountain Blvd., 94619.
Oceanside. Oceanside Photo & Telescope Astro. Soc., 929 Buena Rosa Court, Fallbrook, 92028, Penny Hauck, (619) 723-0684.
Palm Desert. Astro. Soc. of the Desert, College of the Desert, 43-500 Monterey Ave., 92260, Ashley McDermott, (619) 346-8041.
Pasadena. Mount Wilson Observatory Assoc., 813 Santa Barbara St., 91101.
Ramon. Tri-Valley Astron., 20 Cedar Point Loop, Apt. 213, 94583, Alan Gorski.
Ridgecrest. China Lake Astro. Soc., Box 1783, 93556.
Riverside. Riverside Astro. Soc., Box 51222, 92517-2222, 24-hour Hotline. (714)   689-0116.
Rocklin. Sacramento Valley Astro. Soc., Box 575, 95677, R. S. Marasso, (916)   624-3333.
San Bernardino. San Bernardino Amateur Astron., 1345 Garner Ave., 92411, David E. Garcia.
San Diego. San Diego Astro. Assoc., Box 23215, 92193, Greg Cade, (619) 495-1787.
San Francisco. S.T.A.R.S., City College of San Francisco, 50 Phelan Ave., 94112, (415) 239-3242.
San Francisco. San Francisco Amateur Astron., 114 Museum Way, 94114, Bob Levenson, (415) 468-3592.
San Jose. San Jose Astro. Assoc., 3509 Calico Ave., 95124, Jim Van Nuland, (408) 997-3347.
San Juan Bautista. Fremont Peak Observatory Assoc., Box 1110, 95045, Rick Morales, (408) 623-4255.

San Luis Obispo. Central Coast Astro. Soc., Box 1415, 93406, Lee Coombs, (805) 466-2788.

San Marcos. Astro. Soc. of Southern California, Box 2046, 92069, Michael Wartenberg, (619) 741-6128.

San Mateo. San Mateo Astro. Soc., Box 974, Station A, 94403, Bob Bruynesteyn, (415) 349-3743.

Santa Barbara. Santa Barbara Astro. Club, Box 3702, 93130, Clyde Kirkpatrick, (805) 966-3695.

Santa Cruz. Santa Cruz Astro. Club, L-2 San Carlos Ln., 95065, James M. Bricken, (408) 335-2450.

Santa Monica. Santa Monica Amateur Astro. Club, 1415 Michigan Ave., 90404, Robert Lozano, (213) 450-1944.

Santa Rosa. Sonoma County Amateur Astron., Box 183, 95402, Ed Megill, (707) 795-7829.

Simi Valley. Ventura County Astro. Soc., Box 982, 93062, Rick Williams, (805) 523-3066.

Sonora. Omega-One, 21558 American River Dr., 95370, Lance Kearns, (209) 533-2260.

Stockton. Stockton Astro. Soc., Box 243, 95201.

Tulare. Tulare Astro. Assoc., Box 515, 93275, Stanley H. Manro, (209) 685-0585.

Van Nuys. Los Angeles Valley College Astro. Club, 5800 Fulton Ave., 91401, Bruce Dale, (818) 781-1200.

Vandenberg AFB. Vandenberg Amateur Astro. Soc., Box 5321, 93937, Steve Ball, (805) 733-2488.

Walnut Creek. Mt. Diablo Astro. Soc., 1302 Roseann Dr., Martinez, 94553, Ginger Todd, (415) 229-3914.

Whittier. Western Observatorium, 14517 E. Broadway, 90604, Roger Wilcox, (213) 941-9755.

Woodland Hills. Polaris Astro. Assoc., 22018 Ybarra Road, 91364, Ray G. Coutchie, (818) 347-8922.

COLORADO

Colorado Springs. Colorado Springs Astro. Soc., Box 62022, 80962, Jim Lawrence, (719) 593-0801.

Colorado Springs. Rocky Mountain Astrophysical Group, 12815 Porcupine Ln., 80908, Paul Van Slyke, (719) 495-3828.

Denver. Denver Astro. Soc., Box 10814, 80210, Dave Trott, (303) 871-5172.

Grand Junction. Western Colorado Astro. Club, Box 55032, 81505, Harry Brauneis.

Longmont. Longmont Astro. Soc., 2944 Colgate Dr., 80501, James Wilson.

Pueblo. Southern Colorado Astro. Assoc., 2801 8th Ave., 81003, Mike Bosley.

CONNECTICUT

New Haven. Astro. Soc. of New Haven, Box 3005, 06515, Linda Rainey.

Oakdale. Thames Astro. Soc., 55 Doyle Road, 06370, Kathleen Audette, (203) 859-0071.

Stamford. Fairfield County Astro. Soc., Stamford Museum Observatory, 39 Scofieldtown Road, 06903, Charles Scovil, (203) 322-1648.

Stratford. Boothe Astro. Soc., 114 Kings College Place, 06497.

Torrington. Litchfield Hills Amateur Astro. Club, 361 Funston Ave., 06790, D. H. Wheeler, (203) 489-8898.

Waterbury. Mattatuck Astro. Soc., Mattatuck Community College, 750 Chase Parkway, 06708, J. Lawrence Pond, (203) 575-8236.

West Hartford. Astro. Soc. of Greater Hartford, Gengras Planetarium, 950 Trout Brook Dr., 06119, Jay Sottolano, (203) 651-0096.

Westport. Westport Astro. Soc., Box 5118, 06880.

DELAWARE

Wilmington. Delaware Astro. Soc., Box 652, 19899.

Wilmington. Mount Cuba Astro. Observatory Inc., Box 3915, Hillside Hill Road, Greenville, 19807, Grace Quirk, (302) 654-6407.

FLORIDA

Bradenton. Local Group of Deep-Sky Observers, 2311 23rd Ave. West, 34205, Vic & Lynne Menard, (813) 747-8334.

Callahan. Callahan Astro. Soc., Route 3, Box 1062, 32011.

Cocoa. Brevard Astro. Soc., Box 1084, 32923, Tom Olsen, (407) 725-2451.

Fort Lauderdale. South Florida Amateur Astron. Assoc., 16001 West State Road 84, 33326, Phil Hauger, (305) 721-4159.

Fort Pierce. Treasure Coast Astro. Soc., 131 Queen Catherina Court, 34949, Walter Swentzell, (407) 465-4454.

Ft. Myers. South West Florida Astro. Soc., Box 6583, Miracle Mile Station, 33911-6583, Manuel J. Mon, (813) 936-5669.

Jacksonville. Northeast Florida Astro. Soc., Box 16574, 32245-7338, Harold Carney, (904) 645-9310.

Largo. Tampa Area Astro. Soc., 9890 82nd St. N., 34647, John Novak, (813) 392-4699.

Melbourne. FIT Astro. Soc., Florida Institute of Technology, Physics Dept., 32901, Susan Shufelt, (407) 768-8000.

Melbourne. Indian River Astro. Soc., 1201 Sunnypoint Dr., 32935.

Melrose. Alachua Astro. Club, Route 2, Box 2915, 32666, Andrea Vann-Jenson, (904) 475-1014.

Miami. Belen Astro. Club, 500 SW 127th Ave., 33134, Leo Nuûez, (305) 223-8600.

Miami. Southern Cross Astro. Soc., 13841 SW 106 St., 33186, Robert Ward, (305) 385-5021.

Orlando. Central Florida Astro. Soc., 810 East Rollins St., 32803, Doug Moningtan, (407) 896-7151.

Pensacola. Escambia Amateur Astron., 6235 Omie Circle, 32504-7625, J. Wayne Wooten, (904) 477-8859.

St. Augustine. Ancient City Astro. Club, Box 546, 32085-0546, A.L. Ponjee, (904) 797-5704.

St. Petersburg. St. Petersburg Astro. Club Inc., 594 59th St. South, 33707, Daniel W. Bricker, (813) 343-1594.

Tampa. Museum Astro. Resource Soc. (MARS), 2602 E. 98th Ave., 33612, Craig MacDougal, (813) 933-9617.

West Palm Beach. Astro. Soc. of the Palm Beaches, South Florida Science Museum, 4801 Dreher Trail N., 33405, Gary M. Bogner, (407) 832-1988.

GEORGIA

Athens. Athens Astro. Assoc., 160 Plantation Dr., 30605, Maurice E. Snook, (404) 543-3753.

Atlanta. Astro. Soc. of the Atlantic, Center for High Angular Resolution Astro., Georgia State University, 30303, (404) 264-0451.

Atlanta. Atlanta Astro. Club Inc., Box 29631, 30359, Hal Crawford, (404) 320-9156.

Augusta. Astro. Club of Augusta, Box 6373, 30916, Roger Venable, (404) 736-1740.

Macon. Middle Georgia Astro. Soc., Museum of Arts and Sciences, 4182 Forsyth Road, 31210, Gary L. Kretsinger, (912) 956-2145.

Savannah. Oglethorpe Astro. Assoc., 4405 Paulsen St., 31405, Vickie Psillos, (912) 355-6705.

HAWAII

Captain Cook. Mauna Kea Astro. Soc., 84-5095 Hawaii Belt Road, 96704, Howard Yamasaki, (808) 328-9201.
Honolulu. Hawaiian Astro. Soc., Box 17671, 96817, Mike Kaczmarski.
Puunene. Maui Astro. Club, Box 1168, 96784, Daniel O'Connell, (808) 878-3172.

IDAHO

Boise. Boise Astro. Soc., Box 8386, 83707.
Idaho Falls. Idaho Falls Astro. Soc., 1710 Claremont Ln., 83404, Jim **Ruggiero**, (208) 524-6317.
Pocatello. Pocatello Astro. Soc., 1543 Saratoga, 83201.

ILLINOIS

Batavia. Fox Valley Astro. Soc., Box 508, 60510, David Johnson, (708) 466-4811.
Chicago. Chicago Astro. Soc., Box 48504, 60648, Michael Barrett, (312) 725-5618.
Glenview. Skokie Valley Astron., 910 Glenwood Ln., 60025, Gretchen George, (312) 998-1627.
Jacksonville. Central Illinois Astro. Assoc., Illinois College, 62650, Frederick Pilcher.
Libertyville. Lake County Astro. Soc., 603 Dawes St., 60048, Jack Kramer, (708) 362-0959.
Lisle. Illinois Benedictine College Astro. Soc., 5700 College Road, 60532-2099, Robert S. Buday, (312) 969-6550.
Montgomery. Amateur Astron. of Aurora, 69 Sheffield, 60538.
Mount Olive. Mount Olive Astro. Soc., 412 West 4th St. South, 62069.
Naperville. Naperville Astro. Assoc., 205 N. Mill St., 60540, (708) 355-5357, Bob Crowley.
Normal. Twin City Amateur Astron., Box 755, 61761-0755, Sharon MacDonald, (309) 827-4885.
Peoria. Peoria Astro. Soc., 1125 West Lake Ave., 61614, Charles Lamb, (309) 686-7000.
Quincy. Quincy Astro. Soc., 1200 St. Charles Dr., 62301-7160, Kevin Clymer, (217) 222-6535.
Rock Island. Popular Astro. Club, John Deere Planetarium, Augustana College, 61201, Paul Castle, (309) 786-6119.
Rockford. Rockford Amateur Astron. Inc., 6804 Alvina Road, 61101, Barry or Carol Bearman, (815) 962-6540.
Schaumburg. Northwest Suburban Astron., 1417 Harvard Ln., 60193, Detlef Schmidt.
Springfield. Springfield Astro. Soc., 2224 S. 13th St., 62703, Don Walker.
St. Charles. Argonne Astro. Club, 6N106 White Oak Ln., 60175, H. F. DaBoll, (312) 584-1162.

INDIANA

Evansville. Evansville Astro. Soc., Box 3474, 47733, Scott Conner, (812) 424-8639.
Fort Wayne. Fort Wayne Astro. Soc., Box 6004, 46896, Greg O' Neill.
Hammond. Calumet Astro. Soc., 2925 Cleveland St., 46323, Stephen Miller, (219) 845-9174.
Mooresville. Indiana Astro. Soc., 2 Wilson Dr., 46158, J. Philip May, (317) 831-8387.
South Bend. Michiana Astro. Soc., Box 262, 46624, John Greenlee, (219) 255508501.
Warsaw. Warsaw Astro. Soc., RR 8, Box 236, 46580, Jim Tague, (219) 269-1856.

West Lafayette. Wabash Valley Astro. Soc., 2367 Yeager Road #102, 47906, William Annis, (317) 463-3741.

IOWA

Ames. Ames Area Amateur Astron., 1208 Wilson Ave., 50010, David Oesper, (515) 232-8705.
Burlington. Southeastern Iowa Astro. Club, 601 Walnut St., 52601, Jim Blair, (319) 753-2509.
Cedar Rapids. Cedar Amateur Astron., 1513 Parkwood Ln. NE, 52402, Frank Olsen, (319) 393-5758.
Davenport. Quad Cities Astro. Soc., Box 3706, 52806, Wayne Jens, (319) 326-4995.
Des Moines. Des Moines Astro. Soc., 2307 49th St., 50310-2538, C. L. Allen.

KANSAS

Topeka. Northeast Kansas Amateur Astron. League, Box 951, 66601, Mark Cunningham, (913) 271-5269.
Wichita. Kansas Astro. Observers, Box 49013, 67201, Scott Fielding, (316) 838-5816.

KENTUCKY

Fort Wright. Midwestern Astron., 1643 Elder Court, 41011.
Golden Pond. West Kentucky Amateur Astron., Land Between the Lakes, 42211-9001, Jim Boren, (615) 232-6820.
Lexington. Blue Grass Astro. Soc., 1016 Della Dr., 40504.
Louisville. Louisville Astro. Soc., Box 20742, 40450, Chuck Allen, (502) 228-3043.

LOUISIANA

Baton Rouge. Baton Rouge Astro. Soc. Inc., 17324 Gains Mill Ave., 70817, (504) 291-1685.
Shreveport. Shreveport Astro. Soc., 353 Ockley Dr., 71105, Jennie Goodwin, (318) 865-2433.
Westwego. Pontchartrain Astro. Soc. Inc., 948 Ave. E, 70094, Michael Sandras, (504) 340-0256.

MAINE

Fairfield. Central Maine Astron., 4 Osburne St., 04937, John & Alison Meader.
Glenburn. Penobscot Valley Star Gazers, c/o Jeff Churchill, Temple Dr., 04401, Todd Abbotts, (207) 947-3928.
Wells. Astro. Soc. of Northern New England, Box 1238, 04090, Susan Gagne.

MARYLAND

Arnold. Anne Arundel Assoc. of Amateur Astron., Anne Arundel Community College, 101 College Parkway, 21012.
Baltimore. Baltimore Astro. Soc., 601 Light St., 21230, (301) 685-2370.
Bel Air. Harford County Astro. Soc., Box 906, 21014, Richard Hagenston, (301) 836-4155.
Bowie. Bowie Astro. Soc., 12700 Bridle Place, 20715, Fred Espenak.
Cumberland. Cumberland Astro. Club, 350 Bedford St., 21502, G. Frank Simpson, (301) 722-7606.
Greenbelt. Goddard Astro. Club, Goddard Space Flight Center, Code 440.8, 20771, George Gliba, (301) 345-4760.
Greenbelt. National Capital Astron., 7006 Megan Ln., 20770, David & Joan Dunham.
Hagerstown. Tri-State Astron., Washington County Planetarium,

323 Commonwealth Ave., 21740, Jim Taylor, (304) 274-1886.
Westminster. Westminster Astro. Soc., 3481 Salem Bottom Road, 21157, Curtis Roelle, (301) 848-6384.

MASSACHUSETTS
Amherst. Amherst Area Amateur Astron. Assoc., 1403 South East St., 01002, Tom Whitney, (413) 256-6234.
Boston. Amateur Telescope Makers of Boston, c/o 28 Taschereau Blvd., Nashua, New Hampshire, 03062, Marion Hochuli.
Essex. North Shore Amateur Astro. Club, Box 123, 01929, Michael D. McCoy, (508) 768-3494.
North Plymouth. Cape Cod Astro. Soc., Box 1170, 02556, Michael Petrasko, (508) 563-6045.
Norwell. South Shore Astro. Soc., Box 429, Jacobs Ln., 02061, Rolf Egon, (617) 337-2572.
Rutland. Aldrich Astro. Soc., 7 Cheryl Ann Dr., 01543, Russell A. Chaplis, (508) 886-6777.
West Springfield. Springfield Stars Club, 107 Lower Beverly Hills, 01089, John E. Welch, (413) 734-9179.

MICHIGAN
Adrian. Astro. Soc. of Lenawee County, Box 229, Hudson, 49247, Jim Whitehouse, (517) 448-7173.
Alpena. Huron Amateur Astron., 4963 Truckey Road, 49707, Jim Bruton.
Brighton. Friends United Through Astro., 8191 Woodland Shore, Lot #12, 48116, Gary Anderson, (313) 227-9347.
Detroit. Detroit Astro. Soc., 14298 Lauder Ave., 48219, Jack Brisbin, (313) 981-4096.
East Detroit. Warren Astro. Soc., Box 474, 48021.
East Lansing. Capital Area Astro. Club, Abrams Planetarium, Michigan State University, 48824, David Batch, (517) 355-4676.
Hillsdale. Astro. Soc. of Hillsdale County, 500 North Dunes Road, 49242, Pat Calligan, (517) 439-1295.
Holland. Shoreline Amateur Astro. Assoc., A-4679 46th, 49423, Mark Hogsden.
Kalamazoo. Kalamazoo Astro. Soc., Kalamazoo Public Museum, 315 South Rose St., 49007, Eric Schreur, (616) 345-7092.
Lowell. Grand Rapids Amateur Astro. Assoc., 3308 Kissing Rock Road SE, 49331, Evelyn Marron, (616) 897-7065.
Marquette. Marquette Astro. Soc., Box 761, 49855, Brian Halbrook, (906) 228-6351.
Midland. Sunset Astro. Soc., 4300 Washington, 48640, Bill Albe, (517) 835-4142.
Muskegon. Muskegon Astro. Soc., Box 363, 49443, Dale Johnson, (616) 798-7680.
Petersburg. Toledo Astro. Assoc., 8534 Covert Road, 49270, Courtney Earhart, (313) 856-6204.
Plymouth. Astro. Soc. of Michigan, 11422 Waverly, 48170, Michael Best, (313) 459-2378.
Sturgis. Sturgis Astro. Group, 302 Virginia, 49091, Philip Thrasher, (616) 651-8176.
Traverse City. Grand Traverse Astro. Soc., 5999 Secor, 49684, Jerry Dobek, (616) 922-1260.

MINNESOTA
Afton. 3-M Astro. Soc., 14601 55th St. South, 55001.
Brainerd. North Star Astro. Club, 606 South 8th St., 56401, David Starkka, (218) 829-4029.
Duluth. Arrowhead Astro. Soc., Alworth Planetarium of Minnesota-Duluth, 55812, Glenn Langhorst, (218) 726-7129.
New Ulm. Minnesota Valley Amateur Astron., Route 4, Box 15A, 56073, Roger Dier, (507) 359-2488.
St. Paul. Minnesota Astro. Soc., 30 East 10th St., 55101, Max Radloff, (612) 643-4092.

MISSISSIPPI
French Camp. Rainwater Astro. Assoc., French Camp Academy, 39745, James Hill, (601) 547-6970.
Jackson. Jackson Astro. Assoc., 6207 Winthrop Circle, 39206, Walter Rebmann, (601) 982-2317.

MISSOURI
Blue Springs. Astro. Soc. of Kansas City, Box 400, 64013, Gary Pittman, (816) 228-4238.
Columbia. Central Missouri Amateur Astron., 261 Stardust Ln., 65201.
Dawn. North Central Missouri Amateur Astron., Route 1 Box 111, 64638, Jim Jones.
Florissant. McDonnell Douglas Amateur Astro. Club, 3491 Heather Trail Dr., 63031, Richard Melvin, (314) 569-4773.
Fort Leonard Wood. Ozarks Telescope Makers & Amateur Astron., 29 Williams St., 65473, Peter Bessenbruch, (314) 329-3344.
Hannibal. Mark Twain Astro. Soc., 45 Pioneer Trail, 63401, William J. Remillong.
Hazelwood. Rural Astron. of Missouri, 209C Chapel Ridge, 63042.
Riverside. Prairie View Astro. Soc., 931 NW Valley Ln., 64150.
St. Louis. St. Louis Astro. Soc., Route 1, Box 176-A, Silex, 63377, Tom Butler.

NEBRASKA
Lincoln. Prairie Astro. Club, Box 80553, 68501, Lee Thomas, (402) 483-5639.
Omaha. Omaha Astro. Soc., 5025 S 163 St., 68135, Bill Jacobsen.

NEVADA
Elko. Elko Astro. Soc., 550 South 12th #22, 89801, Len Seymour, (702) 738-7916.
Las Vegas. Las Vegas Astro. Soc., Clark County Community College, 3200 East Cheyenne Ave., 89030, John Ruff, (702) 459-8401.
Reno. Astro. Soc. of Nevada, 825 Wilkinson Ave., 89502, Italo Gavazzi, (702) 329-9946.

NEW HAMPSHIRE
Claremont. Springfield Telescope Makers, Chapel Grove, Box 434, 03734, Norman Fredrick, (603) 542-2928.
Keene. Keene Amateur Astron. Inc., 131 Carroll St., 03431, Paul Fisher, (603) 352-6393.

Penacook. New Hampshire Astro. Soc., 22 Center St., 03303, Nelson D. LaClair, (603) 753-9225.

NEW JERSEY

Boonton. Sheep Hill Astro. Assoc., Box 111, 07005, Evert Bono, (201) 335-5990.

Clifton. North Jersey Astro. Group, Box 4021, Allwood Station, 07012, Grace Casalino, (201) 345-STAR.

Cranford. Amateur Astron. Inc., William Miller Sperry Observatory, 1033 Springfield Ave., 07016, George Chaplenko, (908) 709-7520.

Glassboro. Tychonian Observers, Glassboro State College, Dept. of Physical Science, 08028, W. C. Woods, Jr., (609) 863-7348.

High Bridge. New Jersey Astro. Assoc., Box 214, 08829, Ben Cavotta, (201) 638-8500.

Holmdel. STAR Astro. Soc., Box 547, 07733, Doug Krampert.

Jamesburg. Concordia Astro. Club, Box 22, 08831-0022, Eli Drapkin, (609) 655-1422.

Morristown. Morris Museum Astro. Soc., 6 Normandy Heights Road, 07960, Ronald Russo, 201-386-1848.

Orange. Montclair Telescope Club Inc., Box 704, 07050, Roger Salles, (201) 672-2223.

Princeton. Amateur Astron. Assoc. of Princeton, Box 2017, 08540, Greg Mauro, (609) 585-9465.

Toms River. Astro. Soc. of the Toms River Area (ASTRA), Novins Planetarium, Ocean County College, 08754-2001, Erik Zimmerman, (201) 255-0343.

Williamstown. Willingboro Astro. Soc., RD 8, Box 362-A, 08094, Jack Koch, (609) 629-0180.

NEW MEXICO

Alamogordo. Alamogordo Amateur Astron., 1210 Filipino, 88310, Paula Hamilton, (505) 434-0115.

Albuquerque. Albuquerque Astro. Soc., Box 54072, 87153, Bruce Levin, (505) 299-0891.

Clovis. Clovis Astro. Club, 216 Sandzen, 88101.

Hobbs. Lea County Astro. Soc., NMJC Science Building, 5317 Lovington Highway, 88240, V.G. Berner, (505) 392-4510.

Las Cruces. Astro. Soc. of Las Cruces, Box 921, 88004.

NEW YORK

Bellmore. Astro. Soc. of Long Island, 801 Oak St., 11710, Alan Buckley, (516) 221-5914.

Buffalo. Buffalo Astro. Assoc. Inc., Buffalo Museum of Science, Humboldt Parkway, 14211, Doris Koestler, (716) 683-2970.

Clinton. Mohawk Valley Astro. Soc., Box 674-B, Hamilton College, 13323, Richard Somer, (315) 859-4122.

Corning. Elmira-Corning Astro. Soc., 17 West Hazel St., 14830, Marilyn L. O'Connell, (607) 962-5435.

Dewitt. Syracuse Astro. Soc. Inc., 150 Terrace View Road, 13214, Gershon Blackmore, (315) 446-7167.

Frewsburg. Marshal Martz Memorial Astro. Assoc., 176 Robin Hill Road, 14738, Tom Bemus, (716) 386-4566.

Getzville. Assoc. of Observational Astron. of Western New York, 1955 Hopkins Road, 14068, Carl Milazzo, (716) 688-4869.

Huntington. Long Island Astro. Club, 3 Colby Court, Dix Hills, 11746, Richard Pluschau, (516) 499-6599.

Lake Placid. Tri-Lakes Astro. Soc., 100 Sentinel Road, 12946, George R. Viscome.

Lindenhurst. Amateur Observers' Soc. of New York, 707 South 9th St., 11757, Eric Barnes, (516) 957-3713.

Lockport. Lockport Astro. Assoc., 7053 Northview Dr., 14094, Richard Lahrs, (716) 433-4706.

New York. Amateur Astron. Assoc. of New York Inc., 1010 Park Ave., 10028, John Marshall, (212) 535-2922.

Pearl River. Rockland Astro. Club, 110 Pascack Ave., 10965, Allan Green, (914) 735-7303.

Rochester. RIT Astro. Club, Rochester Institute of Technology, Dept. of Physics 08, 14623.

Rochester. Rochester Academy of Science Astro. Club, 105 Wyandover Road, 14616, David Smith, (716) 621-7593.

Schenectady. Albany Area Amateur Astron., 1529 Valencia Road, 12309, Bob Mulford, (518) 374-8744.

Southold. Custer Institute Astro. Club, Box 1204, Main Bayview Road, 11971, Barbara Lebkuecher, (516) 722-3073.

Staten Island. Astro. Soc. of New York City, 153 Arlo Road, 10301, T. W. Hamilton, (718) 727-1967.

Staten Island. Wagner College Astro. Club, 631 Howard Ave., 10301, Paul Albicocco, (317) 390-3341.

Vestal. Kopernik Astro. Soc., Kopernik Observatory, Underwood Road, 13850, Jay Edwards, (607) 748-3685.

Watertown. Northern New York Astro. Soc., Box 106, 13601, Pete Beaumont, (315) 788-2472.

Yonkers. Westchester Astro. Club, Andrus Planetarium Observatory, 511 Warburton Ave., 10701, (914) 963-4550.

NORTH CAROLINA

Albemarle. Stanly County Astro. Soc., Box 1269, 28001, Herbert Hartley, (704) 982-3728.

Asheville. Starfax Assoc., Box 6084, 28816, Timothy Jordan.

Castle Hayne. Cape Fear Astro. Club, 110 Linville Dr., 28429, Paul Petty, (919) 675-2952.

Fayetteville. Astro. Club of Cumberland County, 2308 Colgate Dr., 28304, Allen Faircloth, (919) 485-8515.

Gastonia. Gaston Astro. Club, Box 953, 28054, Larry Benefield, (704) 825-5341.

Goldsboro. Eastern Carolina Astro. Club, 1607 Munroe Ln., 27543, Robert Krieger, (919) 778-3377.

Granite Falls. Catawba Valley Astro. Club, Route 3, Box 472, 28630, Greg Kirby, (704) 396-7656.

Greensboro. Greensboro Astro. Club, The Natural Science Center, 4301 Lawndale Dr., 27408.

Matthews. Charlotte Amateur Astron. Club, 245 Timber Ln., 28105, Gayle Riggsbee, (704) 846-3136.

Raleigh. Raleigh Astro. Club, Box 10643, 27605, Mark Lang, (919)460-7900.

Salisbury. Astro. Soc. of Rowan County, Margaret Woodson Planetarium, 1636 Parkview Circle, 28144, Ellen Trexler, (704) 636-3462.

Statesville. Piedmont Amateur Astron., Route 2, Box 181, Troutman, 28116, Ronnie Sherril, (704) 528-9316.

Trinity. Thomasville Area Astro. Club, Route 3, Box 273, 27370, Arthur Oates, (919) 431-5062.

Winston-Salem. Forsyth Astro. Soc., 504 Gayron Dr., 27105, Stan Griffin, (919) 744-7141.

NORTH DAKOTA

Bismarck. Dakota Astro. Soc., Box 2539, 58502-2539, John R. Wetsch, (701) 256-3620.

Fargo. Moorhead-Fargo Astro. Soc., 1420 8th St. S., 58103, Robert Brummond.

OHIO

Akron. Astro. Club of Akron, 1630 Thornapple Ave., 44301, Mark

Kochheiser, (216) 724-7761.

Chagrin Falls. Chagrin Valley Astro. Soc., Box 11, 44022, Don Himes, (216) 247-5114.

Cincinnati. Cincinnati Astro. Soc., 5497 Washigo Dr., 45230.

Cleveland. Cuyahoga Astro. Assoc., Box 29089, 44129-0089, Dawn Jenkins, (216) 521-5115.

Cleveland. Lewis Astro. Club, 21000 Brookpark Road, M.S. 16-1, 44135, Martin Mayer, (216) 433-8451.

Columbus. Columbus Astro. Soc., 5280 Cape Cod Ln., 43235, Tom Burns, (614) 459-7742.

Dayton. Miami Valley Astro. Soc., 2629 Ridge Ave., 45414, Wade E. Allen, (513) 275-7431.

Lima. Lima Astro. Club, Box 201, 45801, (419) 422-8313.

Lorain. Black River Astro. Soc., 1631 Maple Dr., 44052, Michael Harkey, (216) 288-8556.

Mansfield. Richland Astro. Soc., Box 1118, 44901, Keith Moore, (419) 468-3542.

New Philadelphia. Tucawaras County Amateur Astro. Soc., 1315 E. High Ave., 44663.

Newton Falls. Mahoning Valley Astro. Soc., 1074 SR534 NW, 44444, Edwin Bishop, (216) 742-3616.

Ravenna. Kent Quadrangle Astro. Soc., 7353 State Route 14, Lot 12, 44266.

Rossford. Northwest Ohio Visual Astron., 1018 E. Elmtree Road, 43460, Frank Myers, (419) 666-8566.

South Point. Ohio Valley Astro. Soc., Route 6, Box 350, 45680, Scott Hogsten.

Tiffin. Sandusky Valley Amateur Astro. Club, 650 S. Washington St., 44883, Keith Moore.

Troy. Stillwater Stargazers, Brukner Nature Center, 5995 Horseshoe Bend Rd., 45373, (513) 698-6494.

Wilmot. Wilderness Center Astro. Club, Box 202, 44689, Dave Gill, (216) 833-9399.

OKLAHOMA

Ardmore. Arbuckle Astro. Soc., 713 Ash NW, 73401, Steve Girard, (405) 226-0339.

Bartlesville. Bartlesville Astro. Soc., 225 SE Fenway Place, 74006, Ken Willcox, (918) 333-1966.

Boise City. Cimarron County Star Gazers, Box 278, 73933, Jim Rosebery, (405) 544-2222.

Enid. Northwest Oklahoma Astro. Soc., 1719 Pawnee, 73703, Dan Mathews, (405) 233-5707.

Oklahoma City. Oklahoma City Astro. Club, Box 21221, 73156, Wayne Wyrick, (405) 424-5545.

Purcell. Central Oklahoma Astro. Club, Box 628, 73080.

Sallisaw. Sallisaw Amateur Telescope Club, 109 South Cedar, 74955, Bob Moody, (918) 775-3739.

Tulsa. Astro. Club of Tulsa, 3832 S. Victor Ave., 74105, Newell Pottorf, (918) 742-7577.

OREGON

Bend. Central Oregon Astro. Soc., 17081 Cooper Dr., 97707.

Eugene. Eugene Astro. Soc., Lane ESD Planetarium, Box 2680, 97402, Cuyla Shelton, (503) 345-7748.

Portland. Rose City Astron., Oregon Museum of Science and Industry, 4015 S.W. Canyon Road, 97221, Dale Fenske, (503) 256-1840.

Roseburg. Umpqua Amateur Astron., Umpqua Community College, 3150 West Military Ave., 97470, Paul Morgan, (503) 673-1081.

Veronia. Northwest Astro. Group, 55371 McDonald Road, 97064, Sandy Mikalow, (503) 429-2430.

PENNSYLVANIA

Allentown. Lehigh Valley Amateur Astro. Soc., East Rock Road, 18103, Rod Hatcher, (908) 859-6441.

Beaver. Beaver Valley Astro. Club, 1335 3rd St., #2, 15009, David Yoder, (412) 728-5255.

Clarks Summit. Lackawanna Astro. Soc., 1112 Fairview Road, 18411, John D. Sabia, (717) 586-0789.

Erie. M-31 Astro. Soc., 86 Applewood Ln., 16509.

Glenshaw. Amateur Astron. Assoc. of Pittsburgh, Box 314, 15116, John Holtz, (412) 224-2510.

Haverford. Rittenhouse Astro. Soc., 216 Elbow Ln., 19041, Nancy Blossom, (215) 642-4337.

Kane. Sir Isaac Newton Astro. Soc., 434 Greeves St., 16735, William Kearney.

Kimberton. Delaware Valley Amateur Astron., Box 662, 19442-0662, Marilyn Michalski, (215) 933-0497.

Lewisberry. Astro. Soc. of Harrisburg, Box 356, 17339, (717) 938-6041.

New Kensington. Kiski Astron., 69 Aluminum City Terrace, 15068, Dennis Sopchack, (412) 337-0509.

Penns Park. Bucks-Mont Astro. Assoc., Box 37, 18943, Dick Sivel.

Reading. Berks County Amateur Astro. Soc., Reading School District Planetarium, 1211 Parkside Dr. South, 19611, Linda Sensenig, (215) 375-9062.

York. York County Parks Astro. Soc., 400 Mundis Race Road, 17402, Jeri Jones, (717) 225-3744.

PUERTO RICO

Rio Piedras. Sociedad de Astron. de Puerto Rico Inc., Montevideo 797, Las Americas, 00921, Benito Aponte, (809) 765-0889.

RHODE ISLAND

Newport. Celestial Observers of Rhode Island, 10 Redcross Terrace, 02840, John Hopf, (401) 521-5680.

North Scituate. Skyscrapers Inc., 47 Peep Toad Road, 02857, Conrad Cardano, (401) 828-0702.

SOUTH CAROLINA

Aiken. Aiken Astro. Club, USC-Aiken, Buliding 106, 171 University Parkway, 29801, Richard M. Albert, (803) 663-3105.

Central. Clemson Area Amateur Astron., 1550 Old Seneca Rd., 29630-9597, Dixon Lomax, (803) 656-2197.

Greenville. Roper Mountain Astron., 504 Roper Mountain Road, 29615, Doug Gegen, (803) 288-8595.

Hanahan. Lowcountry Stargazers, 1313 Springhill Road, 29406, John Smith, (803) 797-0146.

Rock Hill. Carolina Stargazers Astro. Club, Museum of York County, 4621 Mt. Gallant Road, 29730, M. Leon Knott, (803) 329-2121.

West Columbia. Midlands Astro. Club, Box 4321, 29171, Luke Rohlfing, (803) 772-5634.

SOUTH DAKOTA

Rapid City. Black Hills Astro. Soc., 4912 Breckenridge Court, 57702, Michael O'Connor, (605) 342-3793.

TENNESSEE

Brentwood. Barnard-Seyfert Astro. Soc., A. J. Dyer Observatory, 1000 Oman Dr., 32027, A. Heiser, (615) 373-4897.

Chattanooga. Barnard Astro. Soc., Box 90042, 37412, F. Randolph Helms, (615) 622-4762.

159

Kingsport. Bays Mountain Amateur Astron., 853 Bays Mountain Park Road, 37663, (615) 229-9447.

Kingsport. Bristol Astro. Club, 824 Hidden Valley Road, 37663, Ken Childress, (615) 239-3638.

Knoxville. Orion Research Forum and Network, Box 50291, 37950, Bob Miles.

Knoxville. Smoky Mountain Astro. Soc., Box 6204, 516 Beaman St., 37914-0204, Rita Fairman, (615) 637-1121.

Manchester. Middle Tennessee Astro. Soc., 1305 Sycamore St., 37355, Donald W. Male, (615) 728-7321.

Memphis. Memphis Astro. Soc., 1229 Pallwood Road, 38122, Richard Moore, (901) 682-2003.

Memphis. Soc. of Low Energy Observers, 4277 Park Forest Dr., 38141, Kathey Nix.

TEXAS

Abilene. Abilene Astro. Soc. Inc., 1109 Highland Ave., 79605, George Schroeder, (915) 677-5713.

Amarillo. Amarillo Astro. Club, 4324 Summit Circle, 79109, Danny Zumbrun.

Austin. Austin Astro. Soc., Box 12831, 78711, Larry Kemp.

Baycliff. JSC Astro. Soc., 4802 Redfish Reef, 77518.

Beaumont. Astro. Soc. of South East Texas, Box 7943, 77726-7943, Jim Heintzleman, (409) 892-2207.

College Station. Assoc. of Amateur Astron., Texas A&M University, Dept. of Physics, 77843-4242, Roger Smith, (409) 845-4179.

Corpus Christi. Corpus Christi Astro. Soc., 3814 Marion St., 78415, Gary Mayer, (512) 852-7643.

Dallas. Texas Astro. Soc., Box 25162, 75225, Sharon Cherry, (214) 358-6982.

Ft. Worth. Ft. Worth Astro. Soc., Box 161715, 76161-1715, Wade Weaver, (817) 731-0804.

Ft. Worth. General Dynamics Astro. Club, 3400 Bryant-Irvin Road, 76109, Ronald W. Evans, (817) 763-6244.

Ft. Worth. Texas Observers, 1501 Montgomery St., 76107, Don Garland, (817) 732-1631.

Houston. Houston Astro. Soc., Box 20332, 77225-0332, John Chauvin, (713) 639-3452.

Houston. Students for the Exploration & Development of Space, University of Houston, 2700 Bay Area Blvd. Box 198, 77058, Carlos Perez, (713) 486-9359.

Lake Jackson. Brazosport Astro. Soc., 400 College Dr., 77566, Steve Lamb, (409) 297-3984.

Lubbock. South Plains Astro. Club, 1920 46th St., 79412-2214, Wayne Lewis, (806) 763-6800.

McAllen. Valley Astro. Soc., 7005 North 31st St., 78504.

Odessa. Permian Basin Astro. Soc., Box 301, 79760, F.A. McElvaney.

San Angelo. San Angelo Amateur Astro. Assoc., Box 60391, 76906, Joseph Lynch, (915) 944-2113.

San Antonio. San Antonio Astro. Assoc., 6427 Thoreau's Way, 78239, Rick Garcia, (512) 654-9784.

Sanger. Denton County Astro. Soc., 225 Green Springs Circle, 76266, John Love, (817) 458-7479.

Stafford. Fort Bend Astro. Club, Box 942, 77477-0942, Dennis Zwicky, (713) 499-4993.

Texas City. College of the Mainland Astro. Club, Division of Natural Science, 77591, John L. Hubisz, (409) 938-4098.

Tyler. Astro. Soc. of East Texas, Hudnall Planetarium, Box 9020, 75711, Mike Mikule, (214) 561-3710.

Victoria. Crossroads Astro. Club, 233 Spur Dr., 77904, Derek Newton, (512) 578-5452.

Waco. Ursa Major Astro. Soc., Baylor University, Box 7316, 76798.

UTAH

Cedar City. Southern Utah Astro. Group, 147 South 300 East, 84720, Brent Sorensen, (801) 586-2759.

Roy. Ogden Astro. Soc., 2336 West 5650 South, 84067, Bob Tillotson, (801) 773-8106.

Salt Lake City. Salt Lake City Astro. Soc., 15 South State St., 84111, (801) 538-2104.

VERMONT

Williston. Vermont Astro. Soc., Box 782, 05495, Frank Pakulski, (802) 985-3269.

VIRGINIA

Bassett. Piedmont Astro. Club, Box 865, 24055, James I. Marshall.

Blacksburg. Astro. Club of Virginia Tech, Virgina Polytechnic Institute, Physics Dept., 24061, John Abbott, (703) 951-3518.

Chesapeake. Back Bay Amateur Astron., 2808 Flag Road, 23323, Glendon L. Howell, (804) 485-4242.

Fredericksburg. Triangulum Astro. Soc., Box 7464, 22404, Al Ventura Jr., (703) 775-2337.

Hampton. Skywatchers, 109 Mill Point Dr., 23669-3534, Marilyn Ogburn, (804) 723-7453.

Harrisonburg. Shenandoah Valley Astro. Club, James Madison University, Dept. of Physics, 22801, Henry W. Leap, (703) 568-6109.

Lynchburg. Lynchburg Astro. Club, 4648 Locksview Road, 24503, Allen Majewski, (804) 384-5616.

Nathalie. Halifax Skywatchers, Box 22, 24577-0022, Joe Lightcap, (804) 349-3700.

Portsmouth. Astro. Soc. of Tidewater, 4205 Faigle Road, 23703, Leonard Scarr, (804) 464-3880.

Portsmouth. Tidewater Community College Astro. Soc., Physics Dept., State Route 135, 23703, Aubrey Hartman, (804) 488-4354.

Reston. Nova Space Soc., 11208 Chestnut Grove Square #5, 22090-5105, Gary Sanger, (703) 435-0019.

Richmond. Richmond Astro. Soc., 4204 Northwich Road, Midlothian, 23112, Tom Bernhardt, (804) 744-1220.

Roanoke. Roanoke Valley Astro. Soc., 3721 Colony Ln. SW, 24018, Gary Close, (703) 989-3474.

Springfield. Northern Virginia Astro. Club, 6121 Rivanna Dr., 22150, Al Schumann, (703) 256-8359.

Virginia Beach. Tidewater Amateur Telescope Makers, 677 Charlecote Dr., 23464, Don Wright, (804) 424-9430.

WASHINGTON

Bellinghman. Whatcom Assoc. of Celestial Observers, 3700 Taylor Ave., 98226, Tom Masterson, (206) 734-5821.

Centralia. Southwest Washington Astro. Soc., 2421 Leisure Ln., 98531, Carl M. Newton, (206) 736-6144.

Keyport. Olympic Astro. Soc., Box 458, 98345, Roger Miller, (206) 698-0381.

Richland. Tri-City Astro. Club, Box 651, 99352, Jerry Johnson, (509) 783-8806.

Seattle. Seattle Astro. Soc., 14454 119th Place NE, Kirkland, 98034, Marjie Kichline, (206) 523-ASTR.

Seattle. Boeing Employees Astro. Soc., The Boeing Company, Box 3707, Mail Stop 4H-58, 98124, Walter Hazen, (206) 657-6488.

Spokane. Spokane Astro. Soc., Box 8114, 99203-8114, Thom Jenkins, (509) 838-1381.

Tacoma. Tacoma Astro. Soc., 7101 Topaz Dr. SW, 98498, (206) 588-9504.

WEST VIRGINIA
Cross Lanes. Kanawha Valley Astro. Soc., 5204 Dellway Dr., 25313, Roger Chapman.
Tridelphia. Oglebay Institute Astro. Assoc., Box 436 Chapel Hill, 26059, Henry Winchester.

WISCONSIN
Beloit. Rock Valley Astro. Soc., 2220 E. Ridge Road, 53511.
Fall Creek. Chippewa Valley Astro. Soc., Hobbs Observatory, Route 2, Box 94, 54742, Bob Elliott, (715) 877-2787.
Franklin. Wehr Astro. Soc., 9701 West College Ave., 53132, Karen Kearns, (414) 425-8550.
Grafton. Northern Cross Science Foundation, 1327 - 11th Ave., 53024, William Fisher, (414) 377-0468.
Green Bay. Neville Museum Astro. Soc., 210 Museum Place, 54303, Donn Quigley, (414) 436-3767.
La Crosse. La Crosse Area Astro. Soc., Box 2041, 54602-2041, Robert Allen, (608) 785-8669.
Madison. Madison Astro. Soc. Inc., 404 Prospect, Waunakee, 53597, Bob Manske, (608) 849-5287.
Milwaukee. Milwaukee Astro. Soc., W248 S7040 Sugar Maple Dr., Waukesha, 53186, Daniel Koehler, (414) 662-2987.
Oshkosh. Northeast Wisconsin Stargazers, 1815 Kienast Ave., 54901, Don Wyman, (414) 233-3083.
Racine. Racine Astro. Soc., Box 085694, 53408, Don Sorenson, (414) 878-2774.
Sheboygan. Sheboygan Astro. Soc., 1712 N. 6th St., 53081.

WYOMING
Cheyenne. Cheyenne Astro. Soc., 3409 Frontier St., 82001, Marcy Flint, (307) 632-2717.

CANADA

ALBERTA
Calgary. RASC - Calgary Centre, Alberta Science Centre, Box 2100, Station M, T2P 2M5.
Edmonton. RASC - Edmonton Centre, 4203 - 106 B Ave., T6A LK7, Mel Rankin.
Fort McMurray. Astro. Soc. of Fort McMurray, 113 Auger Court, T9J 1E5.
Lethbridge. Lethbridge Astro. Soc., Box 1104, T1J 4A2, Rick Ponomar, (403) 381-1332.

BRITISH COLUMBIA
Prince George. Prince George Astro. Soc., College of New Caledonia, 3330-22nd Ave., V2N 1P8, Bob Nelson, (604) 562-2131.
Vancouver. RASC - Vancouver Centre, Gordon Southam Observatory, 1100 Chestnut St., V6J 3J9.
Vernon. Okanagan Astro. Soc., 4100 25th Ave., V1T 1P4, Peter Kuzel, (604) 545-1226.
Victoria. RASC - Victoria Centre, 801 Stanehill Place, RR #1, V8X 3W9, A. Newton, (604) 478-8065.

MANITOBA
Brandon. Westman Astro. Club, 32 Ashgrove Blvd., R7B 1C2, Jeff Harland, (204) 726-8294.
Winnipeg. Manitoba Astro. Club, 190 Rupert Ave., R3B 0N2, Doug Knight, (204) 956-2830.

Winnipeg. RASC - Winnipeg Centre, Box 215, St. James Post Office, R3J 3R4.

NEWFOUNDLAND
Mount Pearl. RASC - St. John's Centre, Box 944, A1N 3C9, Peter Allston.

NOVA SCOTIA
Halifax. RASC - Halifax Centre, 1747 Summer St., B3H 3A6, Douglas Pitcairn, (902) 463-7196.

ONTARIO
Belle River. RASC - Windsor Centre, 453 East Belle River Road, R.R. 2, N0R 1A0, Lorison Durocher, (519) 728-1332.
Kingston. RASC - Kingston Centre, Box 1793, K7L 5J6, Kim Hay, (613) 353-1189.
London. Astro. London, RASC - London Centre, Box 842, Station B, N6A 4Z3, Eric Clinton.
Niagara Falls. RASC - Niagara Centre, Box 241, L2E 6T3, Gregory Saxon, (416) 732-5560.
Ottawa. RASC - Ottawa Centre, P.O Box 6617, Station J, K2A 3Y7.
Scarborough. Metropolitan Toronto Astro. Soc., 154 John Tabor Trail, M1B 2P8, S. Spinney.
Sudbury. Sudbury Astro. Club, 859 Chestnut Crescent, P3A 5B3, Fred Boyer, (705) 560-3265.
Thunder Bay. RASC - Thunder Bay Centre, 545 Parkway Dr., P7A 5C9, Mrs. B. Connell.
Toronto. North York Astro. Assoc., 26 Chryessa Ave., M6N 4T5, Andreas Gada, (416) 761-1798.
Toronto. RASC - Toronto Centre, McLaughlin Planetarium, 100 Queen's Park, M5S 2C6, B. Ralph Chou, (416) 777-4300.
Waterdown. RASC - Hamilton Centre, Box 1223, L0R 2H0, Pauline Wright, (416) 689-5717.
Waterloo. RASC - Kitchener-Waterloo Centre, 114 Westvale Dr., N2T 1J2, Paul Bigelow, (519) 888-7516.
Wyoming. RASC - Sarnia Centre, Box 103, N0N 1T0.

PRINCE EDWARD ISLAND
Summerside. Athena Community Astro. Club, Box 2500, C1N 4L9.

QUEBEC
Montreal. RASC - Le Soc. de Astronomie-Montreal, Casier Postal 206, Ste. Michael, H2A 3L9, Mark Gelinas.
Montreal. RASC - Montreal Centre, Box 1752, Station B, H3B 3L3, Louie Bernstein, (514) 845-2612.
Ste. Foy. RASC - Quebec Centre, Casier Postal 9396, G1V 4B5, (418) 626-9592.

SASKATCHEWAN
Regina. RASC - Regina Centre, Box 6735, S4S 7E6.
Saskatoon. RASC - Saskatoon Centre, Box 317, Sub. Box #6, S7N 0W0, Mike Wesolowski, (306) 373-0137.

Bibliography

ASTRONOMY, Kalmbach Publishing Co., Waukesha, Wisconsin. Founded in 1973, this monthly is the largest English-language astronomy periodical. It contains plentiful information about observing, astrophotography, what is visible in the sky, amateur astronomy events and meetings, and the latest in the science of astronomy.

Berry, Richard. *Build Your Own Telescope.* 276 pp., hardcover. Charles Scribners' Sons, New York, 1985. An introductory manual explaining how to construct six simple telescopes.

Berry, Richard. *Discover the Stars.* 119 pp., paper. Harmony Books, New York, 1987. The former editor of ASTRONOMY introduces naked-eye, binocular, and small telescope observing using twelve all-sky maps and twenty-three close-up maps.

Bok, Bart J., and Priscilla F. Bok. *The Milky Way.* Fifth ed., 356 pp., hardcover. Harvard University Press, Cambridge, Massachusetts, 1981.

Burnham, Robert. *The Star Book.* 17 pp., spiral-bound. AstroMedia and Cambridge University Press, Milwaukee, 1983. Group of seasonal star maps from ASTRONOMY magazine with a brief discussion of what to see.

Burnham, Robert, Jr. *Burnham's Celestial Handbook.* Three vols., 2,138 pp., paper. Dover Publications, New York, 1978. This voluminous compilation of deep-sky objects contains many photographs, charts, and tables.

Cadogan, Peter. *From Quark to Quasar; notes on the scale of the universe.* 183 pp., hardcover. Cambridge University Press, New York, 1985.

Consolmagno, Guy, and Dan M. Davis. *Turn Left at Orion; a hundred night sky objects to see in a small telescope — and how to find them.* 205 pp., hardcover. Cambridge University Press, New York, 1990. A brief description, sketch, and finder chart for one hundred objects of unusual interest to amateur observers.

Covington, Michael A. *Astrophotography for the Amateur.* 168 pp., hardcover. Cambridge University Press, New York, 1985. The best basic introduction to taking astronomical pictures.

Davis, Joel. *Journey to the Center of Our Galaxy; a voyage in space and time.* 335 pp., hardcover. Contemporary Books, Chicago, 1991.

Dickinson, Terence, and Alan Dyer. *The Backyard Astronomer's Guide.* 295 pp., hardcover. Camden House Publishing Co., Camden East, Ontario, 1991. A luxuriously illustrated beginner's guide to the hobby of amateur astronomy.

Eicher, David J. *Beyond the Solar System; 100 best deep-sky objects for amateur astronomers.* 80 pp., paper. AstroMedia, a division of Kalmbach Publishing Co., Waukesha, Wisconsin, 1992. An introduction to the brightest and most spectacular of the sky's nebulae, clusters, and galaxies.

Eicher, David J., and the editors of *Deep Sky* magazine. *Deep Sky Observing with Small Telescopes.* 331 pp., paper. Enslow Publishers, Hillside, New Jersey, 1989. A beginner's manual for observing deep-sky objects with 2-inch to 6-inch telescopes. Contains extensive listings of objects and many photographs and eyepiece sketches made by backyard observers.

Eicher, David J., ed. *Galaxies and the Universe; an observing guide from* Deep Sky *magazine.* 80 pp., paper. Kalmbach Publishing Co., Waukesha, Wisconsin, 1992. A collection of articles on observing galaxies ranging from bright targets like the Andromeda Galaxy to the most extreme challenges for backyard scopes.

Eicher, David J., ed. *Stars and Galaxies; ASTRONOMY's guide to exploring the cosmos.* 200 pp., hardcover. AstroMedia, a division of Kalmbach Publishing Co., Waukesha, Wisconsin, 1992. Hundreds of photos, sketches, and diagrams supplement forty articles centered on unusually rich regions of sky.

Eicher, David J. *The Universe from Your Backyard.* 188 pp., hardcover. Cambridge University Press and AstroMedia, a division of Kalmbach Publishing Co., New York, 1988. This book is a series of republished "Backyard Astronomer" articles from ASTRONOMY magazine. Included in its coverage are forty-six constellations or groups of constellations and 690 deep-sky objects. A three-color map, eyepiece sketches, and color photographs appear for each constellation.

Ferris, Timothy. *Coming of Age in the Milky Way.* 495 pp., hardcover. William Morrow, New York, 1988.

Ferris, Timothy. *Galaxies.* 191 pp., hardcover. Stewart, Tabori & Chang, New York, 1980. A folio-sized photo essay on galaxies with an engagingly written narrative.

Ferris, Timothy. *The Red Limit; the search for the edge of the universe.* 286 pp., paper. Quill, New York, 1983.

Field, George, and Donald Goldsmith. *The Space Telescope; eyes above the atmosphere.* 276 pp., hardcover. Contemporary Books, Chicago, 1989.

Friedman, Herbert. *The Astronomer's Universe; stars, galaxies, and cosmos.* 359 pp., hardcover. W.W. Norton, New York, 1990.

Goldsmith, Donald. *The Astronomers.* 332 pp., hardcover. St. Martin's Press, New York, 1991.

Harwit, Martin. *Cosmic Discovery; the search, scope, and heritage of astronomy.* 334 pp., hardcover. Basic Books, New York, 1981.

Hawking, Stephen W. *A Brief History of Time, from the Big Bang to black holes.* 198 pp., hardcover. Bantam Books, New York, 1988.

Hodge, Paul. *Galaxies.* 174 pp., hardcover. Harvard University Press, Cambridge, Massachusetts, 1986. A revision of the classic introduction to galaxy research written by Harlow Shapley.

Jones, Kenneth Glyn. *Messier's Nebulae and Star Clusters.* Second ed., 427 pp., hardcover. Cambridge University Press, New York, 1991. One of England's foremost amateur astronomers presents descriptions and eyepiece drawings for each of the Messier objects.

Karkoschka, Erich. *The Observer's Sky Atlas.* 130 pp., paper. Springer-Verlag, New York, 1990. A wonderful pocket-sized star atlas showing enough detail to find bright deep-sky objects.

Laustsen, Svend, Claus Madsen, and Richard M. West. *Exploring the Southern Sky; a pictorial atlas from the European Southern Observatory (ESO).* 274 pp., hardcover. Springer-Verlag, New York, 1987. This work consists of a long string of captions packed about a stunning collection of some of the world's finest color astrophotography.

Levy, David H. *The Sky; a User's Guide.* 295 pp., hardcover. Cambridge University Press, New York, 1991. An introduction to the various phenomena visible in the night sky.

Malin, David, and Paul Murdin. *Colours of the Stars.* 198 pp., hardcover. Cambridge University Press, New York, 1984.

McDonough, Thomas R. *The Search for Extraterrestrial Intelligence; listening for life in the cosmos.* 244 pp., hardcover. John Wiley & Sons, New York, 1987.

Murdin, Paul, and Leslie Murdin. *Supernovae.* 185 pp., hardcover. Cambridge University Press, New York, 1985.

The *New Cosmos;* the astronomy of our Galaxy and beyond. 160 pp., paper. Kalmbach Books, Waukesha, Wisconsin, 1992. The best material from ASTRONOMY magazine provides readers with a hard look at the cutting edge of astronomical research, including much on galaxies.

Newton, Jack, and Philip Teece. *The Guide to Amateur Astronomy.* 327 pp., hardcover. Cambridge University Press, New York, 1988. An introduction to what amateur astronomy is all about.

Overbye, Dennis. *Lonely Hearts of the Cosmos; the story of the scientific quest for the secret of the universe.* 438 pp., hardcover. HarperCollins, New York, 1991.

Peltier, Leslie C. *Leslie Peltier's Guide to the Stars.* 185 pp., paper. AstroMedia Corp. and Cambridge University Press, Milwaukee, 1986. A basic introduction to observing stars, planets, and deep-sky objects with binoculars.

Ridpath, Ian, ed. *Norton's 2000.0 Star Atlas and Reference Handbook.* Eighteenth ed., 179 pp. + 16 charts, hardcover. Longman Scientific and John Wiley and Sons, New York, 1989. A classic, *Norton's* provides charts showing stars down to magnitude 6 and an introductory discussion about observing.

Sagan, Carl. *Cosmos.* 365 pp., hardcover. Random House, New York, 1980.

Silk, Joseph. *The Big Bang.* 485 pp., hardcover. W.H. Freeman and Co., New York, 1980.

Sky & Telescope. Sky Publishing Corp., Cambridge, Massachusetts. The oldest astronomy magazine in America, *Sky & Telescope* contains a monthly "Deep Sky Wonders" column written by the experienced observer Walter Scott Houston.

Tirion, Wil. *Sky Atlas 2000.0.* Twenty-six fold-out folio charts, spiral-bound. Cambridge University Press and Sky Publishing Corp., New York, 1981. A large-scale atlas showing 43,000 stars down to magnitude 8 and 2,500 deep-sky objects in color.

Tirion, Wil, Barry Rappaport, and George Lovi. *Uranometria 2000.0.* Two vols., 473 quarto-sized charts, hardcover. Wilmann-Bell, Inc., Richmond, Virginia, 1987-1988. A minutely detailed, large-scale atlas, *Uranometria 2000.0* shows 332,556 stars down to magnitude 9.5 and many thousands of deep-sky objects.

Vehrenberg, Hans. *Atlas of Deep-Sky Splendors.* Fourth ed., 246 pp., hardcover. Treugesell-Verlag and Sky Publishing Corp., Dusseldorf, 1981. A splendid photographic album containing images of hundreds of deep-sky objects all reproduced at the same scale for easy comparison.

Vehrenberg, Hans, and Dieter Blank. *Handbook of the Constellations.* Fifth ed., 197 pp., hardcover. Treugesell-Verlag, Dusseldorf, 1987. A constellation-by-constellation listing of bright stars and deep-sky objects, neatly fitted about a map showing each star group.

Verschuur, Gerrit L. *Interstellar Matters.* 320 pp., hardcover. Springer-Verlag, New York, 1988.

Weinberg, Steven. *The First Three Minutes; a modern view of the origin of the universe.* 188 pp., hardcover. Basic Books, New York, 1977.

163

Index